Human-Centered Tracking for Autonomous Robots

BY

ZAYED

TABLE OF CONTENTS

Chapter 1

Introduction

Research in robotics has led to a widespread adoption of robots for numerous applications, operating in environments such as laboratories, corporate offices, universities etc. In many applications the robot operates along with the humans in a common workspace, where the humans and robots together carry out the needed task. In a museum or tourist attraction, a robot can assist and guide humans to different visiting sites and explain the various facts about these sites. Here the robot also waits and stops when any human leaves his group or does not follow the robot. In this context; a robot can be a teacher, assistant, friend, or guide.. Humans in the decision-making loop of the robot pose a serious problem as the behaviour of the human is hard to model. Therefore, it is not possible to use a simulation tool for benchmarking with the humans replaced by modelled simulated entities. The expectation of a robot in such a context is also hard to objectively define. In the area of cooperative and social robots, interaction between humans and robots has become a point of research interest. Defining the behaviour of the robot is very important and challenging when operating with humans.

In the domain of cooperative and social robotics, the social interaction between robots and humans has become a prominent point of interest. Here the presence of people creates new challenges for motion planning. The major objective of motion planning is to analyse human motion and its trajectory to socially avoid the humans and safely achieve its goal. In this context, a fewer number of hypotheses are known about the behaviour of the humans when the robot is guiding them. The artificial potential field is generally used for motion planning, but this thesis extends the artificial potential field into a social potential field where additional forces are applied for path selection and motion planning. Here a robot chooses a path in the way a human would who is more consciousness of the need to avoid an obstacle and achieve its goal safely.

The thesis aims to make a robotic guide to take visitors on a guided tour to places such as tourist attraction, shopping centres and other, interesting places. The application area of the proposed work is tourist attractions which visitors are interested in exploring and are interested in knowing about the important and historical facts about the places. Currently, guides operate in these areas, who know the different sites of interest and take visitors on a guided tour. A robot should be able to render the same services. The robot needs to operate cooperatively to guide a group of visitors from the starting point to several visit sites in a sequential manner. Here the robot is a leader and works as a multi-visitor tour guide.

While robots are currently used in such environments, the major emphasis of the current research is restricted to interaction with the users. The aim of this thesis is to study the same topic from a navigation perspective, for which a very little research exists. The major challenge studied here is to handle inattention. Modern tours often see visitors receiving important calls and thus it is important for the guide to wait for such visitors. Similarly, visitors opt out temporarily from the group to see a place, stop to buy something, attend children, go to the washroom, get lost enroute, etc. In all such cases, if the visitor is missing for a sufficient length of time, the guide waits for the missing visitors before continuing and the robot must display the same etiquette.

Detection is often lost when observing humans from a mobile robot due to illumination changes. Further, when a sharp turn is taken by the robot, the humans are likely to be outside the field of the view of the camera. In such cases even though the visibility is lost, the robot must not wait for the currently not visible (and presumably missing) visitors, since the lack of visibility is contextual due to the turning of the robot.

The decision making of the robot in such a social context requires a fairly accurate 3D location of the humans. The addition of humans poses a new problem of human tracking as robot and humans both move in real-time. A knowledge of the position of the humans enables navigational decision-making for the robot including intention to go near the human or away from the human, while being generally directed towards a goal. Tracking helps to anticipate the future positions of the robots and humans in real-time for navigation decision making. Tracking is widely performed for different applications, primarily using Particle Filters or Extended Kalman Filters. Particle filtering method is investigated in the area of computer vision since it has an ability to handle non-Gaussian and non-linear problems. A measurement model (likelihood function) is often non-linear due to noise or clutter in the background (Weng et al., 2006), causing the posterior to be non-linear. Effectiveness of the particle filter is very good due to two factors, one is likelihood function which is used to extract visual information from images and the other is proposal distribution, which is a set of discrete samples that will be drawn. Li and Zhang (2004) introduced a proposal for distribution based on the Kalman filter which depends on a learned motion model. In 3D tracking, when any human leaves the group then unnecessary tracker is available for that person which becomes a cause for unusual delay and increases the time computation for the entire tour. Therefore, the tracker is deleted. Thereafter, when the person re-joins the group, the track is re-initialised. The major challenge in tracking is the lack of visibility of the human of the group for a prolonged time. The aim of tracking is to ensure that position estimates are available even if the person is not visible for some time.

A social robot has to achieve its goal by way of the sensor and camera information without any human help. A large number of motion planning algorithms exist in the literature to make a robot avoid static and dynamic obstacles, while making the robot reach its goal. However, when the robot attempts to avoid a human, there is a social interaction between the human and the robot. Humans follow social conventions when avoiding each other. As an example, typically in India a person drifts leftwards on seeing an oncoming person. This behaviour is natural for the humans. A robot must display the same behaviour when operating as a service robot amidst humans. The thesis injects the same social conventions into the commonly used motion planning algorithms, making the robot behaviour socially compliant as per the expectations of the humans.

In this thesis, social potential field-based motion planning algorithm is developed for successfully achieving goal and avoiding obstacles. Artificial Potential Field is a very popular approach used for motion planning and collision avoidance in static environments (Khatib, 1995; Helbing and Molnar, 1999; Reif and Wang, 1999) and this approach works according to virtual forces. Furthermore, geometric approaches are also applicable to formulate a solution for finding the safest path in the environment for the robot motion. The Follow the Gap method is a geometric collision avoidance algorithm which works on the principle of finding the largest gap in the surroundings and moving through that gap (Sezer and Gokasan, 2012). The algorithm is very attractive as it requires very little parameter tuning as compared to the potential field techniques. The improved Follow the Gap method aims to fix one of the primary limitations of the Follow the Gap method that is extremely long trajectories or failure to reach the destination (Demir and Sezer, 2017; Ozdemir and Sezer, 2018). The Vector Field Histogram is another geometric approach and works on the basis of calculating the obstacle density in a robot's surroundings (Borenstein and Koren, 1999). The thesis induces the social navigation preferences of the humans into the working of these motion planning algorithms.

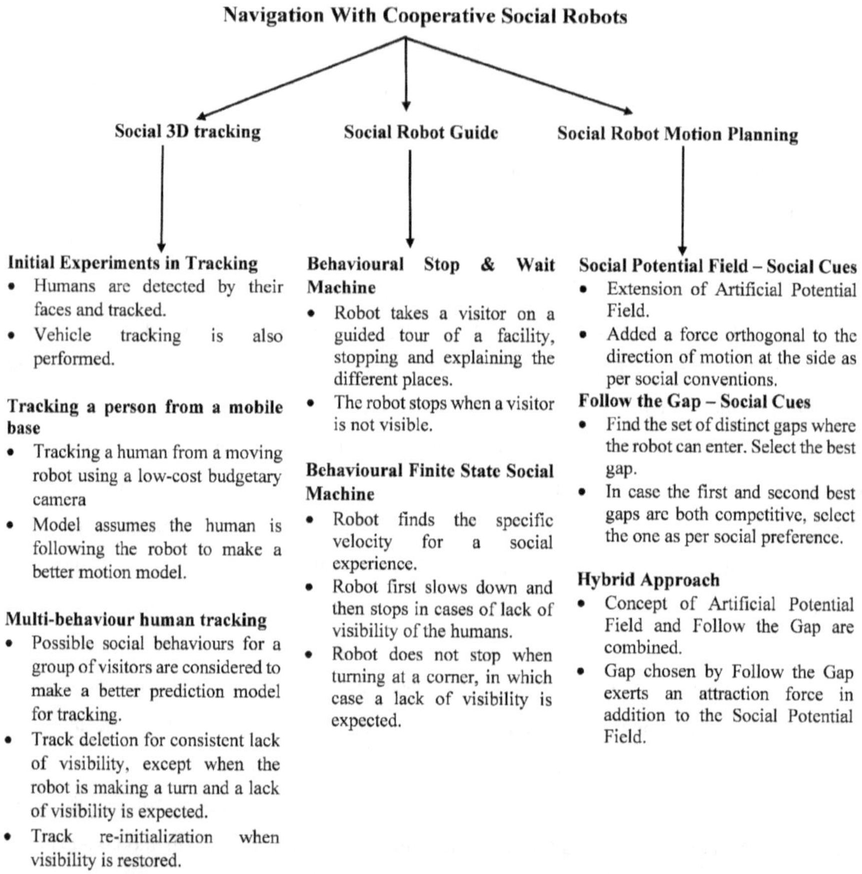

Figure 1.1: Block Diagram of the Thesis

The different algorithms presented in the thesis, together with the major challenges that the algorithms cater to are given in Figure 1.1 and further introduced in the following sections.

1.1 Social Robot Guide

The robot guides a group of visitors from a specified starting point to several labelled sites of visit. To perform this task, it is important for it to see the visitors, which is facilitated by a rear looking camera. When the robot reaches a site of interest, it stops and explains all important facts about that spot, followed by this the robot again proceeds for its next visit site and explains everything about that spot. The robot visits all sites of interest sequentially, waiting for the missing visitors, if needed. The basic working methodology of the approach is shown in Figure 1.2.

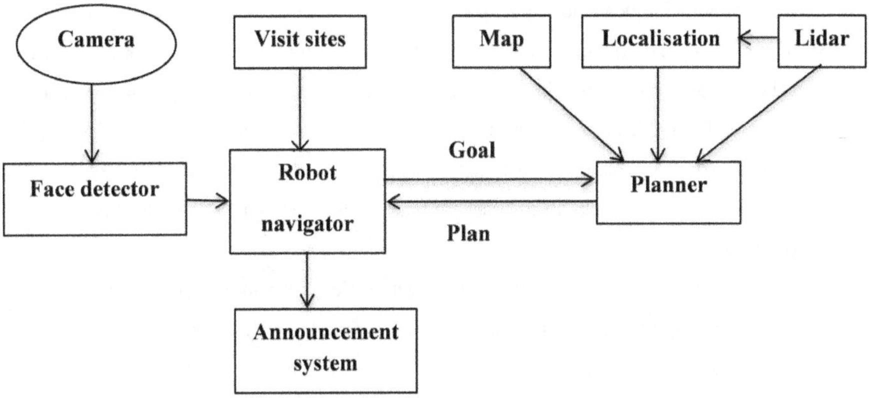

Figure 1.2: Overall Architecture of the system

A Behavioural Finite State Social Machine is modelled to make the robot react to the uncertainties in the detection of the human and to socially take the entire group with it, stopping and announcing if necessary. Furthermore, the robot can anticipate the person moving outside field of view and make decisions accordingly. While the robot is navigating, all the visitors should follow the robot. If any visitor misses or does not follow the robot, then the robot de-accelerates and decreases its velocity from the current velocity to zero. Slowing down gives enough time to the robot to overcome the detection errors due to environment noises. Slowing down gives enough time to the visitors to catch up in case they get temporarily out of view. Once the missing visitor is detected, the robot accelerates and continues the navigation. However, if the missing visitor is not found for a certain length of time, the robot deduces that the visitor is lost or has left the group. In such a case, the robot will make an announcement for the attention of the missing visitors, after the visitors are missing for a sufficient length of time. If the visitors appear and are detected by the robot, then the robot sends an acknowledgement thorough an announcement and continues the tour. The robot moves to the next visit site position, stops there and explains the site to the visitors.

The main objectives of the tour guide application are:

- The robot moves sequentially between visit sites turning at corners to achieve its next goal (visit site) because not every goal may be in the front of the robot as a straight line. The navigation is thus broken down into consecutive sub-goals that ultimately make the robot reach a goal (visit site). There may be additional static and dynamic obstacles on the way to the sub-goal to be reactively avoided.
- The humans may maintain different behaviour in a group such as overtaking the leader robot, switching their positions, overtaking other humans in the group, not following the robot, and stopping at a previous visit site due to their taking a particular interest. The robot must decide its navigation so as to accommodate this behaviour. The robot may choose to slow down to let the humans get back on track or make an announcement for the attention of the missing visitors.
- When the robot makes a turn, the humans get outside the field of view. The robot may be circumventing a narrow corner making it impossible for the humans to place themselves behind the robot immediately. The same case also arises when the robot makes a sharp turn to avoid an obstacle. A simple rule to navigate when all visitors are visible leads to false positives produced by the robot since lack of visibility is because of a lack of space when in fact visitors have not left the group. The system should be aware of such uncertainties and should handle such situations at a corner point while the robot is making a sharp turn. The robot should not slow down or stop.
- Similarly, since the robot and visitors, both are moving, and human detection is not possible at every stage of the journey because of changes in light conditions, lower quality of the camera and a smaller field of view. The robot should not stop unnecessarily when humans are temporarily not visible but are following the robot in the group. The robot should stop and wait only when visitors have actually left the group.

1.2 3D Social Tracking

The theme of 3D tracking is to understand human behaviour in a group while they are following a robot for their navigation. Further, when detection is not possible then also the proposed tracking module can track the people due to a social potential field algorithm. The track is also deleted for a human who leaves its group. The track is re-initialised when the human re-joins the group and is visible to the camera. The 3D world coordinate position is derived from the position in the observed 2D image plane using an inverse camera calibration and an inverse world-to-camera transformation from the image plane. The 3D point is used to re-initialise the tracker. A single low-cost and low field of view camera is required to perform this work.

The objectives of social tracking are:

- To track multiple humans from a moving mobile base
- To track the humans from a limited field of view camera, wherein it is eminent that the humans will often be outside the field of view of the camera.
- To handle the cases when the person temporarily leaves and re-joins the group, while tracking should not happen during this time.
- To accurately predict whether the loss of visibility is due to the person unable to come within the visible region (in which case the tracking must continue) or due to the person leaving the group (in which case the tracking must discontinue).
- To re-initialise tracking when the person re-joins the group.

1.3 Social Robot Motion Planning

Social robot motion planning enables a robot to accomplish its goal and also avoid the obstacles in the path. Using motion planning the robot obtains a collision-free path to achieve its desired goal. In the literature there are various approaches introduced for motion planning for a social robot navigation in different scenarios. The motion planning framework uses a local path planner and a global path planner. The local path planner uses the local information of the environment for generating the immediate action. As an example, the Velocity Obstacle is a local path planner and applicable for avoiding nearby obstacles. The global path planner computes a path according to the goal and the global map. This obtained path is used by the robot to achieve its goal. In this thesis social potential field-based motion planning is applied for robot path planning and human navigation. A global path is broken into sub-goals that are fed into the social-potential based local planner. Because of congested scenarios the robot may have to make a sharp turn in limited spaces and at a corner to achieve the desired goal.

The humans expect the robot to turn left or right as per social conventions while navigating and the human himself or herself drifts towards the opposite side as a social gesture. It is important that the robot keep to such social etiquettes as are expected and displayed by the humans. Further, the robot must adjust as per the human's social gestures of avoiding the robot. While avoiding people the robot should not get very close to them even though this may give a shorter path as its intensions would be unclear. The distance must simultaneously not be very large that is a rude behaviour. The thesis aims to meet all these requirements while designing a motion planning algorithm.

1.4 Motivation

The current works carried out on social robot motion planning has not appreciably looked into the specific behaviour where the humans follow a leader robot, while the research works largely concentrating on socially avoiding the people by a robot or a robot following people. The major challenge that separates this problem from the rest of the variants is that people can show non-cooperative and unexplained behaviour such as leaving the group for a long duration and the robot must still socially adapt to the behaviour. The other major challenge is that there are several corner cases considering the limited field of view of the robot. Unlike other variants, looking at the group at the back, while navigating forward is not an intuitive human behaviour, and therefore the general heuristics of the social motion planning applications fail to recognise the difference in needs in the modelling of this specific behaviour. Consider the example of a robot turning a corner. Now a person behind will be out of view of the camera as the robot makes the turn, and if the robot assumes that the person is missing, it will end up waiting at a sharp corner and thus causing a blockage. Similarly, consider the person briefly leaving the group. Stopping or asking the person to re-join may be impolite.

The social robot applications require a knowledge of the positions of the humans which necessitates tracking the humans in 3D. Here humans may perform different social behaviour such as leaving the group, swapping their position with each other, overtaking each other, unnecessary wait at one place for a certain time, etc. The robot must handle these situations and continuously track the humans without any position loss. This thesis proposes a novel model based on social force and particle filter which can overcome the above-discussed noisy human behaviour and track humans at corners where detection is not possible. In addition, only a less expensive monocular camera (like a webcam) is sufficient to perform 3D tracking in real-time. High expensive cameras (such as stereo cameras, lidar, etc.) are not needed for the proposed model.

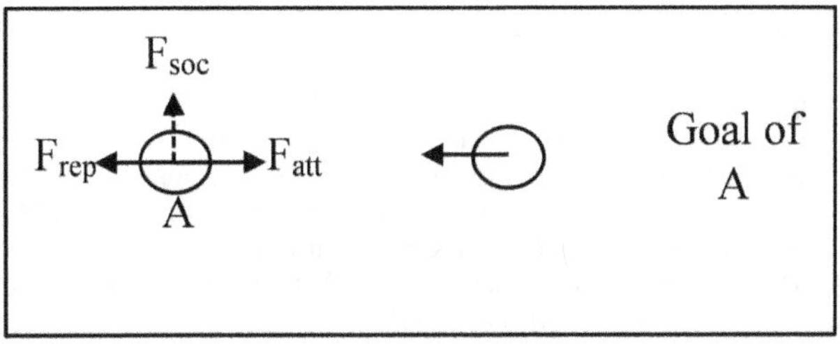

Figure 1.3: Equilibrium condition without the social force when two humans walk opposite to each other

The social robot guide must also socially avoid humans and obstacles on the way for which potential field and geometric approaches are widely used for reactive motion planning. Consider that two humans are facing each other with their goals in the opposite direction and want to pass each other to reach their goals. In a potential field approach, the attraction of the person *A* towards its goal and the repulsion force applied on the person *A* by person *B* will cancel each other out as shown in Figure 1.3. In reality, little errors at the end cause the symmetry to be broken and eventually the humans avoid each other rather than colliding.

However, in such a situation, the human *A* practically sees the other human at the front and starts to change the moving direction much earlier in anticipation to avoid collision. To get the same effect in the robots, an orthogonal force is proposed as shown in Figure 1.3. The same motivation is used to make a better potential field algorithm that is socially compliant. People generally prefer to avoid each other from the left in such situations as per the social conventions in India. Hence the direction of the force is kept leftward as the humans would expect the robot drifting leftward.

A similar convention is also injected in the geometric approaches. The geometric approaches are more proactive in avoiding obstacles; however, the approaches may sometimes cause the robot to be too close to the obstacles. Both potential field and geometric approaches have contrasting pros and cons which motivate the fusion of the two techniques making a hybrid reactive algorithm for the social navigation of the robot (Kala, 2018; Murphy, 2000). This thesis designs a novel hybrid method by combining the force model with the geometric approach in a way that both their characteristics are retained and the social cues are displayed.

1.5 Aims and Objectives

The following are the aims and objectives of the work:

- To track multiple humans from a single monocular camera while the robot and humans both are travelling.
- To develop an ability in the algorithm to delete a track when any human leaves the group and further re-initialize the track when the human re-joins its group.
- To avoid track deletion when the loss of visibility of the human is due to noises.
- To avoid track deletion even if there is a prolonged lack of visibility of the human due to contextual reasons, like the robot is making a sharp turn.
- To model different social behaviours to predict the human motion to enhance tracking, especially when there is a lack of visibility.
- To make a robot that can guide a group of visitors around a facility, making navigation decisions based on a knowledge of the peoples' position.
- To make the robot guide wait for visitors catching up, while to make the robot guide stop for missing visitors, making relevant announcement when necessary.
- To avoid the robot guide slow or slop when the visitors are unlikely to be visible due to contextual reasons like the robot is making a turn.
- To make the robot socially avoid the humans from the correct side and maintaining comforting distances, when planned by using a reactive algorithm.
- To divide motion to goal into different subgoals to reduce the possibility of 'fail to reach goal' due to the robot getting trapped in complex scenarios, enabling motion using a reactive motion planning module.
- To use minimum sensors for implementing all these approaches to make a low-cost budgetary project.

1.6 Main Contributions

The major contribution is for social robot motion planning and its application of 3D tracking for service robotics and navigation of a tour guide service robot. The first contribution is at an application level that is to demonstrate a robot working as a tour guide in a socially compliant way. A social application requires the robot to operate amidst humans and therefore the second contribution is to make a social robot motion planning algorithm. The application also requires an accurate knowledge of the person's position with a limited field of view and therefore the third contribution is the design of a social tracking algorithm. The detailed contributions are as follows:

1.6.1 Contribution 1: Behavioural Finite State Social Machine

The major application is to demonstrate the robot working as a tour guide. This thesis has proposed a Behavioural Finite State Social Machine modelling to make a robot react to the unpredictability of the human-being and to socially take the entire group with it, stopping and announcing if absolutely necessary whose major functionalities are:

i. The major novelty in this approach is that two components are designed, one to navigate the robot adjusting to humans around, and another to make the robot sequentially visit places while telling the visitors about the place. The two components are knit together and synchronised. The robotics problem is solved in a multi-layer hierarchy.

ii. The robot socially responds to the situation so as to mitigate uncertainties in cooperation with the humans. This happens by first slowing down, giving a soft signal to the human; thereafter stopping and subsequently making an announcement, letting the human know the problem.

iii. When the robot makes a sharp turn, it is known that the uncertainties will be large due to the visitors being out of field of view. Therefore, the robot neither slows down nor makes an announcement for the attention of the missing visitors.

iv. The results are demonstrated on the Pioneer LX robot.

1.6.2 Contribution 2: Social robot motion planning algorithm

The social motion planning method assumes a general social convention that the humans prefer to avoid each other from the left and allow aggressive people to overtake by not coming into their way. It is one of the first approaches that directly translates the social human behaviour into the existing motion planning methods from robotics literature. The innovation in the work is as follows:

i. Fusion of a geometric approach which finds the best social gap and a social potential approach for social avoidance of the humans.

ii. Modelling a new social force in the potential field approach, which is applied orthogonal to the vector to the human. The selection of leftward or rightward orthogonal direction vector is carried out based on the social preference.

iii. Using social cues to select a gap for the navigation of the robot, to make the resultant approach socially compliant.

iv. The results are demonstrated under live settings using the Pioneer LX robot.

v. Based on the experimental results, the hybrid approach performs better than the baseline and individual approaches.

1.6.3 Contribution 3: 3D Social Tracking

Furthermore, this thesis also proposes a strategy for 3D social tracking in a real-time environment. The main contributions are:

i. This method studies a new paradigm of social robotics, where multi-human tracking is performed from a budget and limited field of view monocular camera. The challenge lies in humans often getting outside the field of view of the camera.

ii. The system can detect persons leaving the group based on heuristics that incorporate the current context. When any human leaves the group the track is automatically deleted and if the person re-joins, the track is also re-initialised, and tracking continues.

iii. A robot following behaviour is developed that accounts for social attraction and repulsion forces between the people and those with the robot. The model can predict the motion of the person even with a lack of visibility.

iv. The robot can handle the loss of visibility due to sharp turns and corners where detection is not possible attributed to the strong social prediction model and the track is not deleted.

v. The proposed approach more accurately detects people leaving the group and performs a better social tracking when compared to several baseline approaches.

1.7 Visual Results

All experiments have been performed at the Centre of Intelligent Robotics, IIIT Allahabad (Prayagraj) with Pioneer LX and AmigoBot Robots using the Robot Operating System (ROS) framework to ensure the robustness of proposed method in real time. Moreover, the visual outcomes have been published online for each different module which are given as follows:

1) Autonomous Social Robot Navigation using a Behavioural Finite State Social Machine (https://youtu.be/8_o2t-YHuxc)
2) Social Cues in the Autonomous Navigation of Indoor Mobile Robots (https://youtu.be/N8aZrmDhFrI)
3) Trajectory Prediction and Tracking using a Multi-Behaviour Social Particle Filter (https://youtu.be/gSffilRaXn0)
4) Navigation with a Cooperative Social Robot for a group of visitors using face detection and a 'stop and wait' scheme (https://youtu.be/rABqTiqeA48)
5) Social 3D Tracking of Humans from a Mobile Robot for a 'Human Following Robot' Behaviour (https://youtu.be/OzBM7xvAsBE)

1.8 Publications

As a part of this thesis, the following papers are published:

International Journals – SCI/SCIE

1. V. Malviya, A.K. Reddy, R. Kala (2020). Autonomous Social Robot Navigation using a Behavioural Finite State Social Machine. *Robotica*, 38(12), 2266-2289.
2. A.K. Reddy, V. Malviya, R. Kala (2021). Social Cues in the Autonomous Navigation of Indoor Mobile Robots. *International Journal of Social Robotics*, 13, 1335–1358.
3. V. Malviya, R. Kala (2021). Trajectory Prediction and Tracking using a Multi-Behaviour Social Particle Filter. *Applied Intelligence*, DOI: https://doi.org/10.1007/s10489-021-02286-6.
4. V. Malviya, R. Kala. Navigation with a Cooperative Social Robot for a group of visitors using face detection and a 'stop and wait' scheme. (communicated)
5. V. Malviya, R. Kala. Social 3D Tracking of Humans from a Mobile Robot for a 'Human Following Robot' Behaviour. (Communicated)

Conferences

1. V. Malviya, & R. Kala (2018). Tracking Vehicle and Faces: Towards Social Assessment of Human Behaviour. In *2018 Conference on Information and Communication Technology (CICT)*, Jabalpur, India. IEEE, pp. 1-6.

1.9 Organisation of the thesis

The thesis is organised into 6 chapters:

- In chapter 1, a basic introduction of social and service robotics with 3D tracking is given which can be applied for navigating the humans in several places.
- Chapter 2 gives brief details of the background methods which have been utilised in different approaches in the thesis, such as Particle Filter, Kalman Filter, face detection using Convolution Neural Networks, etc.
- Chapter 3 introduces a 3D social tracking algorithm for human following robot behaviour and trajectory prediction by considering different social behaviour of humans in a group. The social behaviour is exploited to give a reasonably accurate tracking even with a loss of observation for intermitted durations of time.
- Chapter 4 provides an application for a robot as a social guide that takes the visitors on a guided tour around several places. The robot slows down on a lack of visibility of the

visitors to allow the visitors to cope up, while the robot stops when the people are not visible for a prolonged time. When the robot makes a turn at a corner, it loses visibility and thus is prevented from stopping.

- Chapter 5 explains a social robot motion planning algorithm with induction of Indian navigation etiquettes and human collision avoidance based on social cues. Two near-colliding humans expect each other to act in a certain socially compliant way, and hence a robot must obey the same social conventions that is demonstrated in this chapter.
- Chapter 6 gives the overall conclusion and future direction.

Chapter 2

Tools and Techniques Used

In the development of social navigation and tracking, a few methods are taken from the literature, and these are extended for further research and development. Human faces are detected for navigation, while the Kalman filter and Particle filter are utilised for tracking and position estimation. The potential field algorithm is used to make a motion model for the filters as well as for the navigation of the robot. This chapter discusses these methodologies.

The chapter is organized as follows. First a Convolution Neural Network is explained in section 2.1, using which the face detection pipeline used in this thesis is explained in section 2.2. The two basic tracking approaches of Kalman Filter and Particle Filter are explained in sections 2.3 and 2.4 respectively. The two basic motion planning techniques of Artificial Potential Field and its extension Social Potential Field are explained in sections 2.5 and 2.6 respectively. The thesis uses Behaviour Finite State Social Machine as a technique to model and program the complex behaviours, which is explained in section 2.7. The thesis also uses the ROS framework that is briefed in section 2.8.

2.1 Convolution Neural Network

CNN is quite a powerful tool in understanding data like images. A typical CNN contains multiple layers, where each layer receives input from the previous layers. First comes the **Input** layer which revives input, the size of this layer varies according to the input data. This input can be understood as a n-dimensional matrix. The **Convolutional** layer applies convolutional matrix filters on the input and generates a consolidated representation; we can understand it as a convolution of the input matrix with a feature-map matrix, where the feature-map matrix comprises of weights that are learnt in the training process. There can be several such layers in a network such as **Pooling** layers in between the convolutional layers which sample out the interesting data points from the receptive field of the previous layers. For example, it can calculate the maximum, minimum or average of the data points. The maximum pooling is a common choice. Each layer neuron is activated based on the activation function. **ReLU** is the most commonly used. The function has non-linearity to imitate non-linear decision boundaries, while the function avoids the problem of the vanishing gradient where only the layers near the output layers learn as the gradient keeps reducing as one goes from the output layer to the input layer. However, based on different use cases this can be changed. The different layers can use different activation functions. This thesis has used the CNN for human face detection in real time by the robot.

2.2 Face detection

The model proposed by Zhang et al. (2016) is applied to continuously detect and localise the face. The face detection model is sub-divided into three fragments stages which are P-Net (fast-proposal network), R-Net (refinement network), and O-Net (output network). Here the candidate window is generated by fast-proposal network, refinement is carried out by way of a refinement network, and a facial landmark and bounding box are obtained from the output network.

Suppose an image is available. By the image pyramid model, it is re-sized into distinct scale and it works as the input for these three fragment stages of the model.

Stage I: Initially a P-Net (fast-proposal network) is applied to generate the candidate (facial) window and bounding box via a regression vector. Furthermore NMS (non-maximum suppression) is employed to integrate the extremely overlapped candidate.

Stage II: In this stage each candidate window is fed into an R-Net (refine network) so that a large number of false positives are eliminated. A bounding box is regressed followed by a non-maximum suppression.

Stage III: This stage is mostly the same as stage I, but the main objective of this new stage is to determine the entire covered area by the human's face with extremely high recognition. The output of this stage is the 'position of five facial landmarks'.

Various experiments and measurements are carried out to determine the performance measurements. Since in progressive CNN (Li et al., 2015) some filters with minor diversity have few restrictions, so in the comparison with multi-class object detection approach, human face detection approach is highly challenging for binary taxonomy. At the end the statistics of each filter are minimised to 3×3 from 5×5 to reduce computational time, however depth is increased to obtain a high performance. Here PReLU (He et al., 2015) works as a non-linear activation function that is applied after convolution to all convolution layers excluding the output layers. ReLU is also called as the most recent development of the deep network (Krizhevsky et al., 2012). Figure 2.1 explains the overall depiction (Zhang et al., 2016) of the convolution neural network with P-Net, R-Net, and O-Net which is designed for human face detection. Here 2 and 1 are used as a step size for pooling and convolution and *MP* and *Conv* are abbreviations for max pooling and convolution.

Cross entropy loss is employed for every sample s_k in the detection phase. Mathematically it can be written as Equation (2.1);

$$loss_k^{det} = -(y_k^{det}\log(p_k) + (1 - y_k^{det})(1 - \log(p_k))) \tag{2.1}$$

Here ground truth label is represented as y_k^{det} and network obtained probability is p_k. It implies that sample being face has a probability p_k.

In the sub-problem of regression on the bounding box, the network offset is determined for each candidate among it and its closest ground truth (top, left, height and width of every bounding box). The main objective of this approach is known as a problem of the regression. Moreover, Euclidean loss is applied for each sample s_k as shown in Equation (2.2);

$$loss_k^{box} = \left\|\hat{y}_k^{box} - y_k^{box}\right\|_2^2 \tag{2.2}$$

Here the term $\hat{y}_k^{box}$ represents the regression output which is obtained from the network module and y_k^{box} is the term which represents the ground-truth coordinate that includes (top, left, width and height) and it implies $y_k^{box} \in R^4$.

The facial landmark step minimises the Euclidean loss, and this problem is similar to the problem of bounding box regression as given by Equation (2.3);

$$loss_k^{landmark} = \left\|\hat{y}_k^{landmark} - y_k^{landmark}\right\|_2^2 \tag{2.3}$$

Here the term $\hat{y}_k^{landmark}$ denotes facial landmark coordinates (computed from the network module) and $y_k^{landmark}$ represents the ground-truth coordinate for k^{th} sample. Here left mouth corner, nose, left eye, right mouth corner, and right eye are the five facial landmarks, each associated with an X and Y coordinate value, and thus $y_k^{landmark} \in R^{10}$.

In the learning process of CNN different kinds of images are used like non-face images, face images, and moderately aligned face image. If the sample belongs to the background, then $loss_k^{det}$ is calculated, while other remaining losses are set to zero. It can be implemented directly using a type indicator function. The target of its learning process is given in Equation (2.4);

$$min \sum_{k=1}^{M'} \sum_{j\in\{det,box,landmark\}} \aleph_j \beth_k^j loss_k^j \tag{2.4}$$

Here M' is the number of training samples, task importance is represented by parameter $\aleph_j$, and $\beth_k^j$ is the sample type indicator.

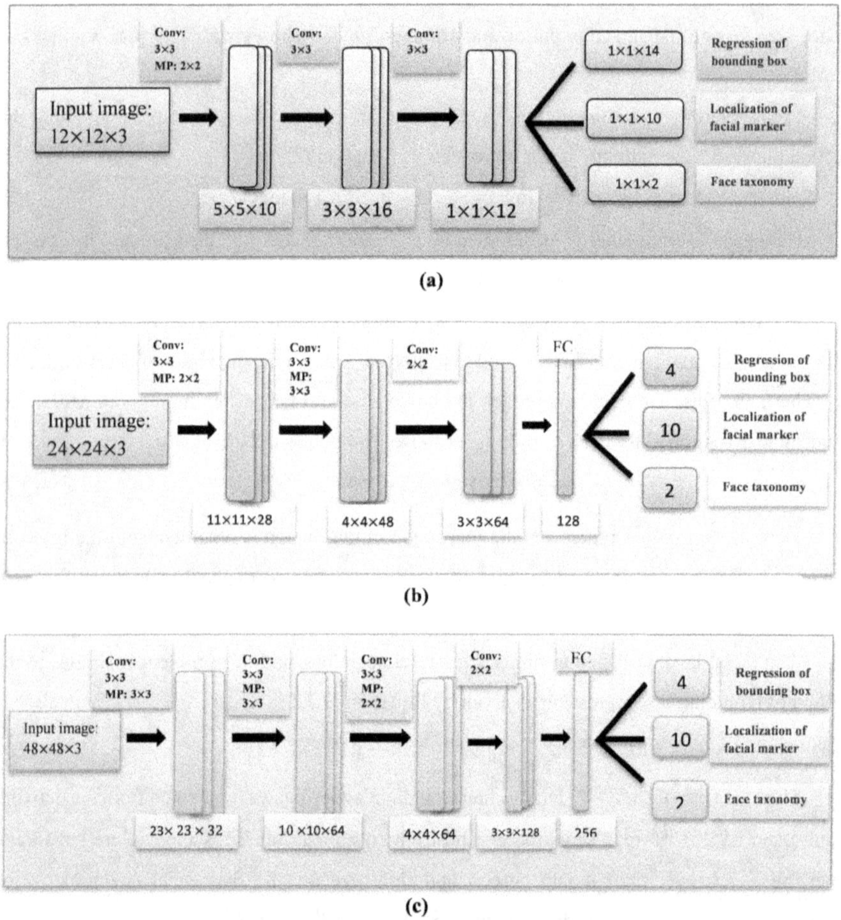

Figure 2.1: Convolution Neural Network (a) Depiction of Proposal Network **(b)** Depiction of Refine Network **(c)** Depiction of Output Network

2.3 Kalman Filter

The Kalman filter is an algorithm that uses a series of mathematical equations for the estimation of the position and velocity of the moving object. The major steps are in the Kalman filter algorithm are: (i) Initialisation, (ii) Prediction, (iii) Computation of the Kalman gain, and (iv) Correction.

The basic design of the Kalman filter is explained in the Figure 2.2. Assuming, that the state at time t-1 is known as X_{t-1}, the object is subjected to motion given by Equation (2.5);

$$X_t = AX_{t-1} + B\mu_t + \omega_t \tag{2.5}$$

Here μ_t is the control variable matrix at time t. Whenever the object is freely moving in a space then gravitation force, acceleration by external force, etc. act on that object that move the object that constitutes the control inputs. ω_t is the noise at time t in the predicted state matrix. In general, this noise matrix cannot be determined. A and B are the transition matrices that are used to convert current state and control input to the new state matrix and the entity of A and B matrix depend on the state matrix and the control variable matrix respectively. Here X_{t-1} is the state matrix at time t-1. The state matrix typically contains position and velocity to track.

In a tracking environment, a measurement (Z_t) is made by relevant sensors that can be observed. If X_t is the system state at time t, the observation made is given by Equation (2.6);

$$Z_t = (H \times X_t) + S_t \tag{2.6}$$

In Equation (2.6), the matrix H relates the system state X_t to the observation Z_t and S_t is the error in the observation (such as error in layer, line or sensor). The error S_t cannot be determined.

We first make a prediction about the state. Let X_t be the estimated state at time t. Assuming that the noise ω_t is from a Gaussian distribution with 0 mean, the most expected predicted state at time t is given by Equation (2.7).

$$X_{t,pred} = AX_{t-1} + B\mu_t \tag{2.7}$$

In the Kalman filter, an error is accumulated in the tracking process. This error is also included in the modelling of the Kalman filter which is modelled as a process covariance matrix. The process covariance matrix is updated by Equation (2.8);

$$P_{t,pred} = AP_{t-1}A^T + Q_t \tag{2.8}$$

Here Q_t is the covariance associated with the noise model and P_{t-1} is the estimated covariance matrix at the previous time t-1.

Followed by this, Kalman gain is computed to correct the predicted position (Equation (2.9)). Kalman gain is used to determine how much observation is to be applied to update the predicted state. Kalman gain is the relative weight between the measurement and the current t state estimate. Intuitively, the value of Kalman gain is lies between 0 to 1, although more formally the term denotes a matrix whose elements lie between 0 and 1. The denominator is thus more formally the inverse.

$$K_t = \frac{er_{pred}}{(er_{pred} + er_{obs})} \tag{2.9}$$

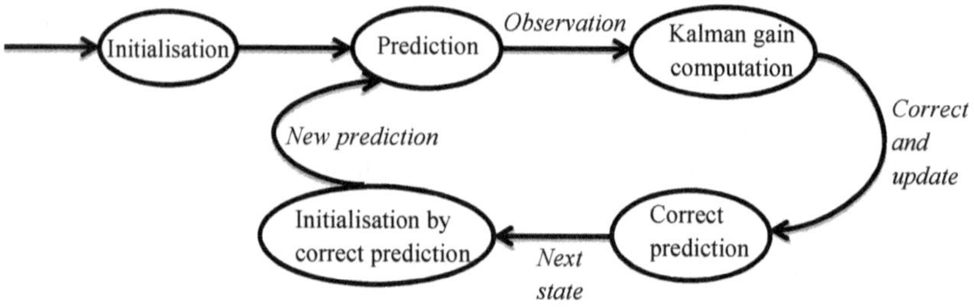

Figure 2.2 Basic design of the Kalman filter

Now $P_{t,pred}$ is the predicted covariance matrix. Here Kalman gain is expressed by Equation (2.10);

$$K_t = \frac{HP_{t,pred}H^T}{HP_{t,pred}H^T + R} \tag{2.10}$$

Here R is sensor noise covariance matrix corresponding to the observation noise model S_t. The update is the state using the Kalman gain is given by Equation (2.11);

$$X_t = X_{t,pred} + (K_t \times (Z_t - H \times X_{t,pred})) \tag{2.11}$$

Here Z_t is the observation at time t and X_t is the position estimate after observation.

As the new position is predicted and corrected in the continuation of the tracking process, it is also important to update the process covariance matrix which is carried out using the Kalman gain. The new process covariance is determined by Equation (2.12);

$$P_t = (I - (K_t \times H)) \times P_{t-1,pred} \tag{2.12}$$

Here I is the identity matrix and $P_{t-1,pred}$ is the process covariance matrix one step before. The overall architecture of the Kalman filter in the form of a flow diagram has been represented in Figure 2.3.

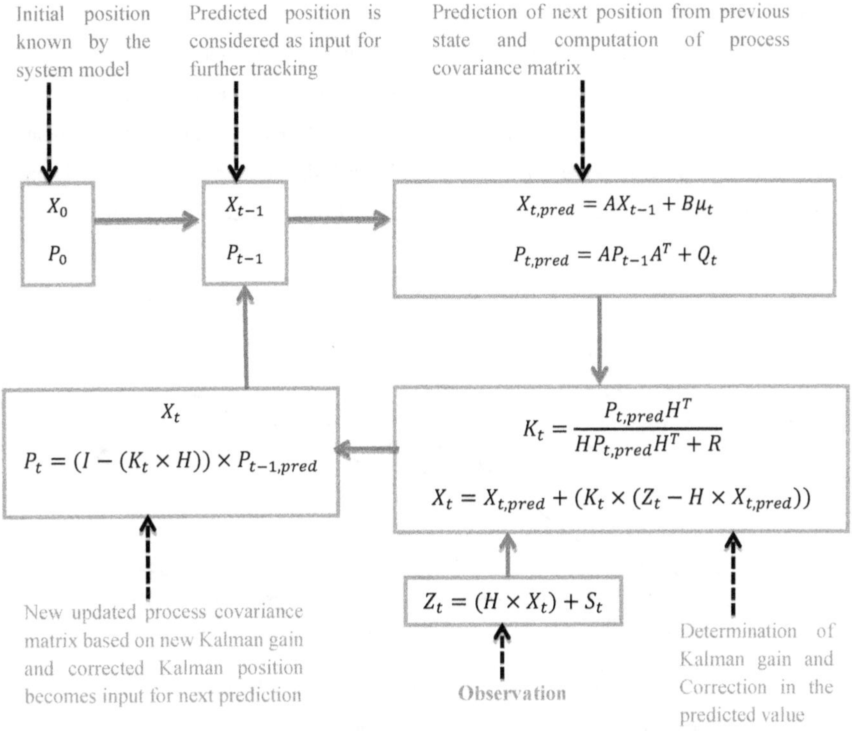

Figure 2.3: Workflow diagram of Kalman filter

The limitation of the Kalman Filter is that the motion and observation models are both linear that may not be practically possible for many real-life systems. The Extended Kalman Filter (EKF) extend the Kalman Filter for human motion prediction using a non-linear motion and observation model. The EKF locally linearises the motion and observation models around the current operating point (state) by using a Jacobian matrix to compute the first order derivative of the motion and observation models. The EKF uses the Taylor series expansion of the non-linear space and uses the first order derivative term only neglecting the higher order derivative terms to get a linear space.

2.4 Particle Filter

3D particle filter is used for multi-human tracking. Here a set of particles or samples is used to represent the probability density function. Each particle represents the state variable with some numerical value. It is an efficient and alternative method to demonstrate and keep the probability density function non-stationary and non-linear. The basic design of the particle filter is pictured in Figure 2.4. Here the robot tracks humans who are moving in a group. Let the position of the human be given by Eq. (2.13);

$$s = \binom{x}{y} \tag{2.13}$$

Here (x, y) is the 2D coordinate of the person.

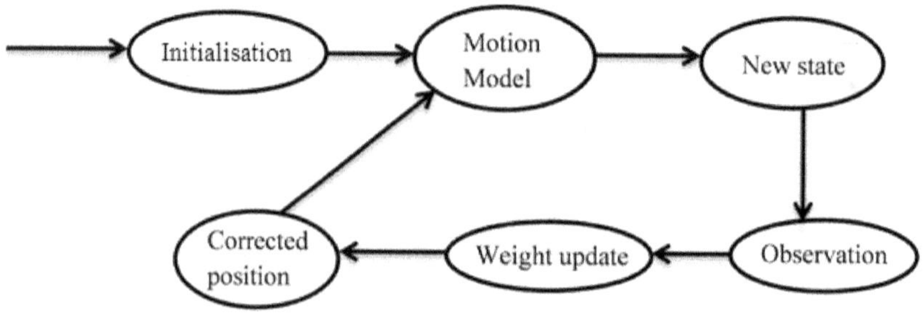

Figure 2.4 Basic design of the Particle filter

In the particle filter (Choi and Christensen, 2012; Choi and Christensen, 2011), the posterior density function $P(h_{t,p}|\mathfrak{Z}_{1:t}, S_{1:t})$ is given by all the previous observations $\mathfrak{Z}_{1:t}$ and an unknown control signal which is ultimately derived from every previous position of the robot ($R_{1:t}$) and other people ($q_{1:t,p}$), together making the system state $S_{1:t} = \{q_{1:t,p}, R_{1:t}\}$. The subscript t is for time and p is for the person being tracked. The particle filter is defined (in Eq. (2.14)) as a set of weighted particles.

$$\mathfrak{K}_{t,p} = \left\{\left(h_{t,p}^{(1)}, w_{t,p}^{(1)}\right), \left(h_{t,p}^{(2)}, w_{t,p}^{(2)}\right), \dots, \left(h_{t,p}^{(n)}, w_{t,p}^{(n)}\right)\right\} \tag{2.14}$$

In Eq. (2.14), $h_{t,p}^{(i)}$ represents the ith particle sample which is the position or state of the human p at time t. Here subscript t is used for a time, p for the person being tracked, and n denotes the number of the particles. $w_{t,p}^{(1)}$ is the corresponding weight of the particle.

The first step in the implementation of a particle filter is to initialise the particles based on the known state of the system. In this case, it is assumed that the initial position of the persons is known and therefore the particles are spread around the known initial positions with some noise. The initial weights are assumed to be the same for all particles. The filter continuously updates the positions and weights of the particles.

The filter applies the motion model to update the knowledge of the state of the system. Our 3D motion model is based on multiple social forces and anticipates the position of every human in the group. The problem is, given the positions of all people and the robot, the system must anticipate the control input that the person will invoke given by Eq. (2.15) and the update state of the person because of the control given by Eq. (2.16).

$$\mathcal{U}_{t,p}^{(i)} = anticipate\left(h_{t,p}^{(i)}, S_t\right) + \mathcal{Y}_{noise} \tag{2.15}$$

$$h_{t+1,p}^{(i)} = \mathcal{k}\left(h_{t,p}, \mathcal{U}_{t,p}\right) \tag{2.16}$$

Together Eq. (2.15-2.16) represents the motion model of every particle. In the above expression, $\mathcal{U}_{t,p}^{(i)}$ is the anticipated control for humans which consists of the linear and angular speeds. Here function "*anticipate*" tries to capture humans' motion on a given current state $h_{t,p}^{(i)}$ corresponding to the ith particle. Moreover, $\mathcal{Y}_{noise}$ is the motion noise used for uncertainty in model anticipation and $\mathcal{k}$ is the kinematic equation of motion.

The particle filter then uses an observation $\mathcal{Z}_{t,c(o)}$. Here *c* is the correspondence function that maps the *o*th observation to the corresponding person *p*=*c*(*o*) and *t* is the time. The observation is used to weight the particles. The weight ($\mathrm{w}_{t,p}^{(i)}$) is directly proportional to the likelihood function $p\left(\mathcal{Z}_{t,c(o)} | h_{t,p}^{(\mathrm{i})}\right)$. This likelihood function is the probability of observing $\mathcal{Z}_{t,p}$, given the current state as $h_{t,p}^{(i)}$.

The implementation of such a particle filter will soon have many particles in poor areas. Hence a re-sampling is carried out where the particles in good areas with high weights are replicated many times, while the ones at poor areas with low weight values die off, representing a new particle distribution with more particles at good areas.

In our proposed model of the 3D particle filter, the current state $q_{t,p}$ of the human *p* is anticipated by the weighted mean of the particle as expressed in Eq. (2.17);

$$q_{t,p} = \sum_{i=1}^{n} w_{t,p}^{(i)} h_{t,p}^{(i)} \tag{2.17}$$

2.5 Artificial Potential Field

Artificial Potential Field (APF) is a motion planning algorithm the purpose of which is for the robot to achieve the desired goal and avoid obstacles. This algorithm is based on the attraction and repulsion force towards the goal and obstacle respectively. In this phenomenon the robot is assumed to be highly negatively charged (artificially) and other moving agent and obstacles have also a negative charge. Further, the goal is assumed to be positively charged. Here the negative charges will repel each-other, and the positively charged goal will attract the negatively charged robot. The modelling of attraction and repulsion force is based on the distance between the robot to goal and robot to obstacle. The attraction force towards the goal is given by Equation (2.18),

$$F_{attr}^{Goal} = -k_{att}(r(x,y) - goal(x,y)) \tag{2.18}$$

Here k_{att} is the attraction constant, $r(x,y)$ is current position of the robot, and $goal(x,y)$ the goal position in two dimensions.

There are some limitations to in Equation (2.18) such as when the robot is very far away from the goal then the attraction force applied over the robot will be high and as the result, the robot will move with very high speed and may collide with the obstacles or the boundary of the workspace. Additionally, when the goal is near to the robot then Equation (2.18) is a better modelling as it makes the robot stop at the goal and this modelling is termed as a quadratic modelling. It is necessary to set some threshold value over the attraction force so that attraction force should be some constant when the robot is far away from the goal, known as the conic model. A combination of quadratic and conic model of the attraction force is given by Equation (2.19),

$$F_{attr}^{Goal} = \begin{cases} -k_{attr}(r(x,y) - goal(x,y)), & d(r,goal) \le d_{attr} \\ d_{attr}k_{attr}\frac{(r(x,y)-goal(x,y))}{d(r,goal)}, & d(r,goal) > d_{attr} \end{cases} \tag{2.19}$$

The second part of the Equation (2.19) is the conic modelling of the attraction force whereas first part is quadratic. Both modellings are necessary because if only conic modelling is applied then the robot will not stop at the goal and will always oscillate on it, and further when the goal is very near then the robot velocity should be decreasing in nature to stop at the goal but with the conic model the robot will continuously move with a constant velocity and will never stop. The quadratic force only on the other hand will become very large far away from the goal. In Equation (2.19) d_{attr} is the distance threshold value and $\frac{(r(x,y)-goal(x,y))}{d(r,goal)}$ is the unit vector of the attraction force to determine the direction of the force.

Similarly, repulsion force modelling of the artificial potential field is given by Equation (2.20),

$$F_{rep}^{o} = \begin{cases} k_{rep}\left(\frac{1}{d(r,o)} - \frac{1}{d_{rep}}\right)\frac{(r(x,y)-o(x,y))}{d^3(r,o)}, & d(r,o) \le d_{rep} \\ 0, & d(r,o) > d_{rep} \end{cases} \tag{2.20}$$

In Equation (2.20), d_{rep} is the repulsive threshold distance and $d(r,o)$ is the distance between robot and obstacle (o). Furthermore, when robot is far away from the obstacle then it is not necessary to apply the repulsion force therefore it is set as zero as represented in the second part of the Equation (2.20).

Hence total force of the APF is given by Equation (2.21),

$$F_{total} = F_{attr}^{Goal} + \Sigma_o F_{rep}^{o} \tag{2.21}$$

The robot moves with the application of this virtual force that constantly moves the robot towards the goal, avoiding obstacles if the obstacles are in the vicinity.

2.6 Social Potential Field

Social Potential Field (SPF) is an extension of the APF. APF consists only of an attraction force towards the goal and repulsion from the obstacles; but SPF also consists of other forces including these attraction and repulsion forces. When humans are moving in a group then attraction and repulsion can also happen among the humans due to their friendship and enmity, this is known as group behaviour. This attraction and repulsion force applied by the person p is known as the social force ($F_{social,p}$). In APF the obstacles only repel the robot, asking the robot to be far away from the obstacles as much as possible. However, while moving in a group with peers, humans always keep a respectable social distance. Being far away from the peers is interpreted as rude behaviour. Similarly, being too close is interpreted as a breach of personal space that is also discouraged. The comfortable distance is typically a measure of the acquaintance level of the peers. Two people who do not know each other will keep a large distance while friends will prefer to keep a much smaller social distance between each other. Enemies can often avoid each other maintaining a much higher distance. The force is attractive when the distances between the closest peers is large, while repulsive if the distances are very small, leading to the comfortable social distances being at an equilibrium. Therefore, the total force of the SPF is the sum of F_{social} and all forces (F_{attr}^{Goal}, F_{rep}^{obs}) calculated from the APF as given in the following Equation (2.22),

$$F_{total}^{soc} = \Sigma_p F_{social,p} + F_{attr}^{Goal} + \Sigma_{obs} F_{rep}^{obs} \tag{2.22}$$

2.7 Behaviour Finite State Machine

The Behaviour Finite State Machine (BFSM) integrates the theme and prominent factor of agent behaviour. Every state of the BFSM makes a control rule for goal seeking, characteristic of the agent, possible mood of the agent, etc. (Curtis et al., 2015; Ulicny and Thalmann, 2002; Bandini et al., 2006). BFSM is different from a Finite State Machine (FSM). In an FSM, a state denotes the condition at a particular instance of time as encoded in the state. However, in BFSM, the state denotes a control rule that is used to move the agent. The motion of the agent causes a change in its pose and velocity that is not possible in FSM. More concretely, the BFSM is a type of hybrid-FSM where every state consists of some discrete state variables that do not change (here the control rule name) and some continuous variables that may change within the state (here the pose and velocity of the agent).

A transition function of the BFSM is dependent on changes in the agent behaviour. In the preceding state of the BFSM, a Boolean test is performed. This condition test consists of an arbitrary logic scheme to achieve the target. The target element finds the different state of the BFSM where the agent automatically moves after satisfying the condition element. An Action element enables changing the state of the BFSM by the agent. The Action element is only executed when the agent enters into the BFSM state domain. Generally, BFSM is used for robot motion planning and its movement related task such as navigation, fault tolerance, etc. Thus, the robot changes its state or behaviour multiple times in the navigation process. The change occurs as the transition function finds an invoking logic that triggers a change in state and hence the invoked behaviour.

The robotic guide is a complex software where the agent needs to display multiple behaviours at different times. At times it may be waiting for the visitors, at other times it may be explaining the visit site, at other times it may be moving towards a visit site, while sometimes it may be slowing down while being uncertain about a visitor. The robot changes its behaviours based on its pose and the estimated status of the visitors. The BFSM gives the ability to enlist all the behaviours and the transition conditions that trigger a change in the displayed behaviours that makes it easy to design the software.

2.8 Autonomous mobile robot navigation using ROS

Robot Operating System is an open-source framework for building robot applications using a combination of different software library, algorithms, and developer tools. Basically, Robot

Operating System (ROS) is not an operating system but a meta operating system where there in an underlying operating system to assist the ROS. Ideally, ROS uses C++ and Python languages for the programming perspective and it works on Linux, MacOS, and Windows 10. Initially, it was released in 2007. Moreover, ROS also provides several tools and libraries to develop the proposed algorithm on a real robot and its application, that include device drivers, message-passing, visualizers, hardware-abstraction, package management, etc. In this thesis, experiments were initially performed on a simulator followed by the real robot.

2.8.1 ROS Terminologies

A ROS system is represented by a graph, where collection of programs are the nodes of a graph which are executable file within ROS package, and here one program that communicates with the other using a ROS client library are simply connected by edges. Furthermore, nodes can also utilise and provide a service. Two nodes communicate and can transfer the data and information with each other through a topic which is a name for a stream of messages that can have different types. ROS message is used for describing a data value via a simplified message description language and these message descriptions are saved in *.msg* file extension in the subdirectory of the ROS package. ROS library functions provides a Publisher-Subscriber Interface for passing a message in ROS. Here any ROS node can be a Publisher or Subscriber where Publisher holds and passes the message to a specific topic and Subscriber subscribes the same topic which is published by the Publisher. Publisher and Subscriber can publish and subscribe more than one topic, although the awareness of each-other's existence is not known by Publisher and Subscribers.

Apparently, node can be deeply concealed in a different directory hierarchy, and to communicate and automate these different node '*rosrun*' as a command-line utility is used which is provided by the ROS. This command allows to use a package name without knowing the package name to run a node since it is applicable to search a package for the program and also passes the supplied parameter. Additionally, '*roslaunch*' is a command-line tool which is included in ROS for starting and launching a collection of nodes. Moreover, *roslaunch* works on *.launch* file extension that are XML files to describe a collection of nodes along with different topic remapping and parameters. There is another command known as '*rosservice*' to listen and query a ROS service that also contains a Python library for retrieving information about different services and dynamically invoking them. This command takes the subcommand 'call', service name and its argument.

In a ROS, actionlib is a package that provides a standardised interface for interfacing with preemptible task and programs, its specification is defined by a .action file extension that has a goal definition, result definition, result definition, and feedback definition.

To start ROS '*roscore*' command is used on terminator that provides a connection information to the nodes so that messages among different nodes can be transmitted. Followed by other set of commands needs to run on another terminator (or terminator).

2.8.2 Using manufacturer provided libraries with ROS

The robot is programmed using different libraries provided by the manufacturer available as packages. The first package is the "*rosarnl*" that contains a "*rosarnl_node*" to provide ROS interface for the ARNL features such as robot navigation and localisation system for robot platforms. ARNL is a manufacturer-specific library used to control the robot. Here common messages are defined by the *nav_msgs* which is used to interact with the navigation stack. The dependency of *ros-arnl* are *geometry_msgs*, *actionlib*, *nav_msgs* etc. Here goal can be cancelled or sent via *actionlib* from the ROS navigation stack. The following command is to be typed onto terminator to run *rosarnl_node*:

rosrun roaarnl rosarnl_node

The above command will connect to the robot. ARNL is accessed and configured via a '*MobileEyes*' software, a manufacturer-specific software for controlling the robot using a GUI tool. *MobileEyes* is used for map scanning, and localization as shown in Figure 2.5.

For map loading and to trigger initial localization, the following command should be entered on a separate terminator.

rosservice call /rosarnl_node/load_map_file /path/file_name.map

eg: rosservice call /rosarnl_node/load_map_file /home/vaibhav/Desktop/Robita_final.map

rosservice call /rosarnl_node/global_localization

The first command uses a *rosservice* to load the map from a local file. The map can be made by using manufacturer-provided libraries and laved on a local file that is loaded here. The second command uses a *rosservice* to perform global localization using the map file.

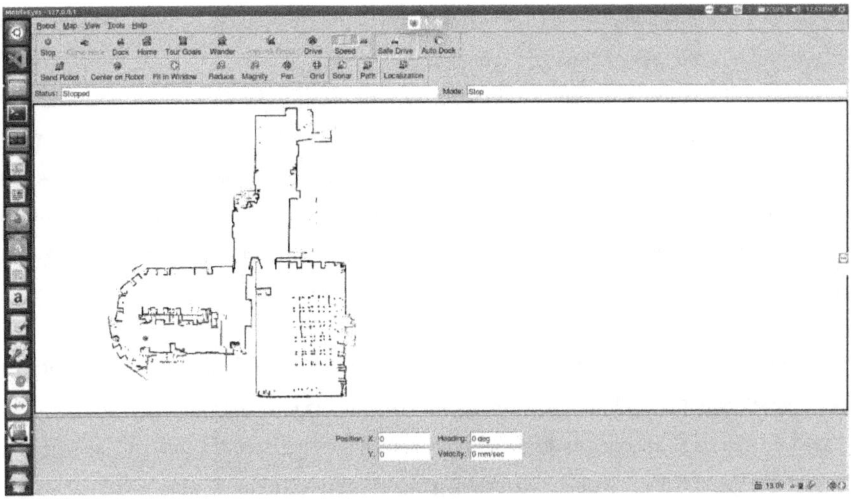

Figure 2.5 MobileEyes

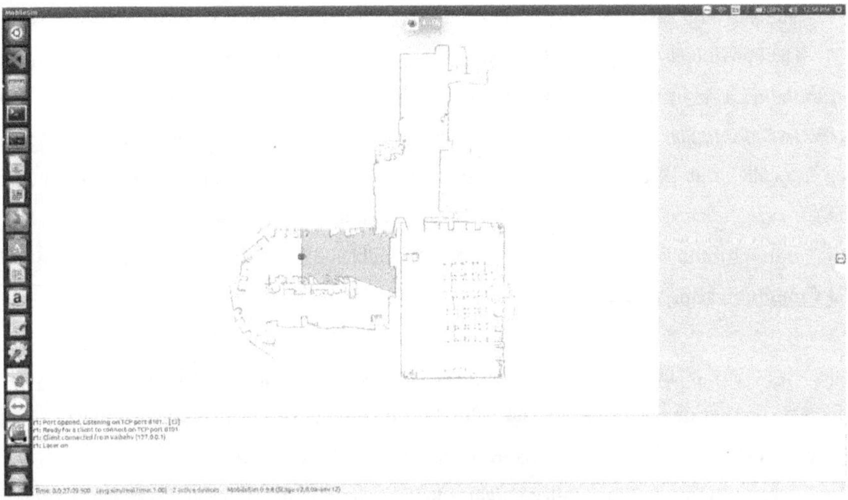

Figure 2.6 MobileSim

Furthermore, *MobileSim* and *rosaria* are also used for the development of the Behavioural Finite State Social Machine algorithm presented in this thesis. *MobileSim* is a simulator (as shown in Figure 2.6) to perform experiment. The simulator can be used in place of a real robot and provides the same functionalities as a real robot. The environment map can be loaded on this simulator with the following instruction command:

MobileSim m /path/file_name.map -r p3dx:robot_variable_name –start x_position,y_position,0.0

In the above command *MobilSim* is a command-line prompt to run the *MobileSim* simulator, -*m* is used to load a *.map* file followed by the path of the *.map* file with file name of the map is given in the command line. *-r p3dx* is used to represent robot model, and *variable_name* is the name of the robot to call in the program followed by the starting position given as an input where the robot will be initially situated.

eg: MobileSim -m /home/vaibahv/Downloads/New_robita_lab.map -r p3dx:r1 –start 6000,6000,0.00

Here *RosAria* is used that provides a ROS interface for several mobile robot. Robot velocity, acceleration, other information from robot base is implemented via a *RosAria* node. The package provides ROS interface to the *Aria* library, a manufacturer specific library to interact with the robot. The general instruction to run rosaria package is given below:

rosrun rosaria RosAria __name:= robot_variable_name _port:=localhost:8101 _publish_aria_lasers:=true

The above command is used to run a *rosaria* package with *RosAria* node. Here name is used to define a robot with the same variable name as used in the *MobileSim* simulator. When *rosaria* and *MobileSim* are running on the same computer then *rosaria* can detect the *MobileSim* using a localhost as IP and a default port of 8101, both specified using the *~port* parameter. *_publish_aria_lasers* is a parameter specified to state that the lidar sensor should be invoked. E.g.

rosrun rosaria RosAria __name:=r1 _port:=localhost:8101 _publish_aria_lasers:=true

2.8.3 Communicating with a server

Master ROS has been used in this thesis for better computation time, where robot can be connected by any other device (high configuration laptop). The aim is to let nodes on several systems communicate to each other, such that the compute at different systems could be used, while the programs exchange messages using topics as if all the topics were available on a local system. Master ROS uses a centralized architecture, where one of the machines acts as a central server called as the Master ROS, identified using a ROS_MASTER_URI and ROS_IP. Here ROS_IP is the environment variable that sets the network address of the central server whereas ROS_MASTER_URI is another important variable that explains a node where the ROS Master exists. The steps for Master ROS connection between a high-configuration server (Master ROS) and a low-configuration robot are given below:

1. Run '*roscore*' on any device (high configuration server):

   vaibahv@server:~$ roscore

2. Run two other commands (given below) on a different terminator on the same device where '*roscore*' is already running.

   vaibahv@server:~$ export ROS_MASTER_URI=http://172.16.21.200:11311

   vaibahv@server:~$ export ROS_IP=172.16.21.200

   Here 172.16.21.200 is the IP address of the server and 11311 is the default port of ROS.

3. Because the robot does not have a monitor/keyboard, a remote login needs to be done. Run a *ssh* command on a terminator to access the robot. Assume 172.16.21.202 is the robot's IP address and guest is the username on the robot's machine.

   vaibahv@server:~$ ssh guest@172.16.21.202

4. After successfully login on this robot system with credential, whatever command will be typed that will directly run on the robot. First the robot needs to be told about the Master ROS (server), done by the following commands. The IP and port are the same as that of the server.

   guest@robot:~$ export ROS_MASTER_URI=http://172.16.21.200:11311

   guest@robot:~$ export ROS_IP=172.16.21.200

5. Let us assume that there are several programs or standard libraries downloaded as source that are together placed in a workspace folder named 'catkin_ws'. The

   guest@robot:~$ cd catkin_ws/

   guest@robot:~/catkin_ws$. devel/setup.bash

   apparently, in the current directory, 'build' and 'devel' folder exists. There are several setup.*sh file exist in the sided the 'devel'folder. Therefore, sourcing any of these files will resultantly overlay current workspace on the top of environment which are using. So, before continuing source the new setup.*sh file; this command is used.

6. To run *rosaria* package on the robot, type the following command onto the same terminator where robot has been accessed. Here port */dev/ttyUSB0* specifies that *rosaria* is executed on a physical robot.

guest@robot:~/catkin_ws$ rosrun rosaria RosAria __name:=r1 _port:=/dev/ttyUSB0 _publish_aria_lasers:=true

2.8.4 An Example of ROS-Graph

This thesis uses lots of publishers and subscribers for transferring the data as a communication signal among different programs, such as visitor availability, visit site status, velocity of the robot, audio file that will be played by the robot when the robot accomplishes a goal or any visitor misses their goal, etc. Figure 2.7, elaborates the ROS-Graph of Behavioural Finite State Social Machine, an algorithm developed in this thesis. In Figure 2.7, it is shown how nodes and topics are connected to each-other and transfer the information. Here */r1*, */robot1* and */sequencer* are the nodes (programs) which are denoted inside an elliptical shape. The topics such as '*/r1/pose*', '*/r1/cmd_vel*', '*/angle_errror*', etc. are denoted by a rectangular shape. Here the arrows are used to represent the information flow as messages from Publisher to Subscriber. Figure 2.7 shows how different nodes (programs) run independently and pass information as messages to each other, using which the entire algorithm can be programmed as separate nodes.

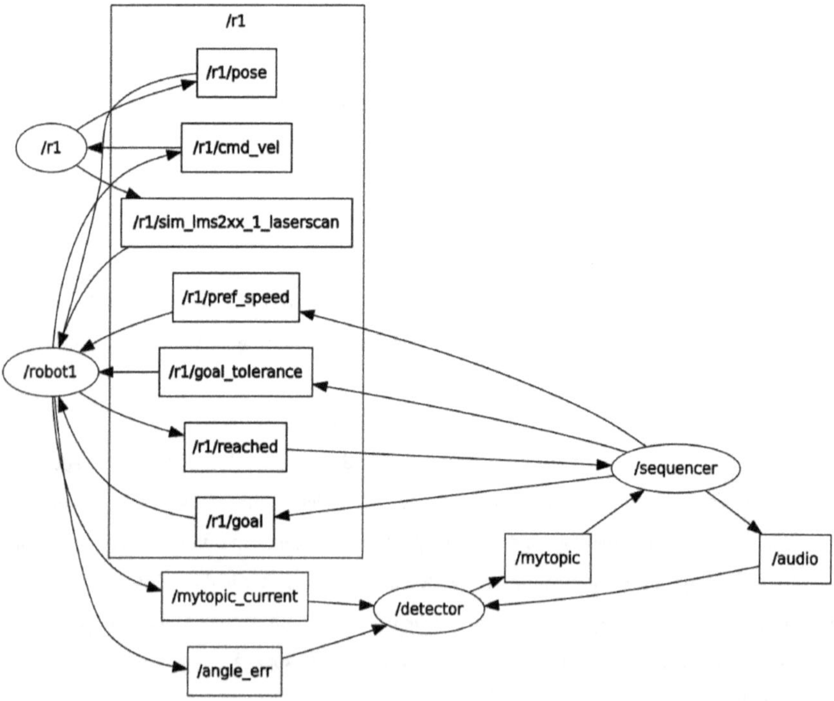

Figure 2.7 An example of ROS-Graph

Chapter 3

Social Tracking

3.1 Introduction

The increasing possibility and opportunity of the social robots and deep learning has become a cause to develop robots that operate with humans. The common scenarios correspond to human following the robot, robot following another robot, robot moving towards a goal with a human following the robot, or a robotic convoy guiding a human group. In the last case, the trajectory of a leader robot is determined by the different patterns of the robots and humans.

3.1.1 Tracking using a Stationary Camera

Many researchers have proposed various algorithms to track human motion in order to understand their behaviour. It has a wide spectrum of applications such as surveillance, ability to automatically monitor human activity, behavioural analysis, video conferencing, etc. These systems are applicable in security sensitive areas such as border, airports, lobbies of building, and areas of military and police. Human motion tracking has become a very challenging task because people get occluded by others, different lighting conditions, complex environments, cluttered background, colour distribution of the target people and background colour model are different in dark and bright environments.

At the initial stage, to track a human motion, this thesis utilises colour information while tracking from a single and stationary camera. The humans primarily operate in two very distinct types of environments. The first is highly structured environments consisting of queues, a fixed path of motion and rule-based display of behaviour. The best example is the traffic scenario operating outdoors. The second is the highly unstructured environments wherein a human operates in wide open spaces with few rules. The best example is navigating in wide open office spaces. To get the best understanding of diverse human behaviour, first this thesis targets both the extreme cases of an outdoor and unstructured indoor office environment. The key performance indicator is the tracking error of the trajectory with different heights of the stationary camera and different distances between the tracked object and the robot.

3.1.2 Tracking from a Mobile Robot using the Human Following Robot behaviour

It is impractical for social robots to use external fixed cameras. Hence, this thesis also develops a framework named as Human Following Robot behaviour (HFR) for human tracking for applications where the human is following the robot. Here a social particle filter framework is developed to track the human who is following the robot. Estimation of human position in a dynamic system with noisy measurements is a very active area of research in computer vision and social robot navigation. The linear velocity and angular velocity of the robot may also change drastically which becomes a cause for sharp deviation into the human' velocity and position. The case of robot making a sharp turn is particularly interesting. If tracking in the image plane, a sharp robot turn produces a very sharp change in the position of the person in the image, which is an effect that most conventional trackers do not account for. Most conventional tracking algorithms smooth such behaviours which leads to erroneous results. Hence the thesis proposes a mechanism to track the human in 3D real world coordinates, while making a social motion model that accurately predicts the behaviour of the human, knowing the motion made by the robot.

The current applications rely on expensive social robots that come with a lot of autonomy. However, if robots are to be widely used, it is important for them to be affordable. Attributed to advances in deep learning, it is possible to use low end vision cameras to reach a goal while avoiding obstacles. Further, current research makes it feasible to localise the robot using affordable odometry systems with special markers for long term correction. This thesis hence concentrates on the problem of tracking and localising a moving human. This thesis demonstrates the tracking on a budget differential wheel drive mobile robot using a single limited field of view monocular camera for tracking the human in 3D. Once the face of the human is detected at a certain position, then the proposed particle filter tracker initialises on the human face and tracks the human during the motion. The robot initial position is assumed to be the origin $(0, 0, 0)$ and also the human is assumed to be standing behind the robot initially. As the robot moves, the human follows the robot by maintaining a certain social distance.

The particle filter tracker predicts the next position, as both the human and robots move, assuming a social motion model. The social motion model assumes that the human always goes towards the robot maintaining a social distance. The human further has a smooth transformation of speed at any point of time including switching between behaviour which is incorporated into the motion model. The social motion model is important since the human will be out of the visibility of the robot for prolonged times. The conventional trackers fail in such a situation. Even if parameter tweaking is carried out to let tracking continue, the uncertainty grows extremely large to make tracking useless. However, in real life, a mother leading a child does not very often look back and always has an indicative idea of the position of the child. This motivates a tracker that can have a robust performance even with an extremely limited observation, which is facilitated by a knowledge of the social behaviour of the human when following the robot. Different kinds of fusion between social and random motion models are also tried, however experiments reveal that the social model is itself extremely robust and gives the best results.

The observation model uses the monocular camera mounted on the rear of the robot. To determine the error between the predicted position and the actual position, the 3D world coordinate system (particle) is converted into a 2D image plane coordinate. To convert the 3D real world into a 2D image plane, first the predicted human position is transformed into the coordinate system of the camera of the robot (knowing the position of the robot) and then the calibration matrix is used to project the face onto the 2D image plane. The 2D coordinate in the image plane represents the centre position of the human face in the image plane. Further, the error is determined between the observed position in the image and the predicted position of the human face using this technique.

The key performance indicator is the difference between the actual trajectory of the human and the trajectory predicted from the Particle Filter model on a dataset with the following challenges:

i. The camera is budgetary that is often affected by environment noises
ii. The robot and humans often circumvent narrow corners leading to a loss of visibility
iii. Humans may get outside the field of view of the camera for prolong times
iv. The humans move as per the discussed social norms which is already incorporated in the database

3.1.3 Multi-behaviour tracking of a group

The thesis further extends the work to multiple people in a group. In the proposed application, the humans in a social group also follow the robot while having an affinity towards each other, which makes the tracking hard. The motion module of the tracker may predict that the humans continuously move towards the robot, however, it may not be possible due to a larger number of people needing more personal space. Similarly, people often form groups within themselves and move as a cluster. In this chapter, the multi-behaviour 3D particle social force model is proposed to solve this problem which includes various social forces such as attraction and repulsion between human to human and between humans to the robot.

During the navigation process of the robot, the human can also behave erratically, they can interchange their position with someone in a group, one human can overtake another one, and it is a possibility that human may go ahead of the robot while following it. Specifically, one additional behaviour can occur that humans do not follow the robot and still stand in their previous position to look at a place for a longer time. These social aspects are hard to model as their triggering conditions are non-observable. The tracker should be able to handle such situations still.

Often, the human can get a telephone call and to attend it, the human must leave the group. If one human leaves the group, another human may also leave the group due to their relationship to each other. Often it happens that a human can leave the group permanently because they might get an emergency call. The proposed model detects such situations and deletes the tracker if any human leaves the group and as the human re-joins his group, the tracker is automatically determined by computing the visitors' position in the 3D real word and tracking them forward.

In the literature, the problem of deleting tracks is solved largely by the time for which the person is not visible. If a human is not detected for a prolonged time, it is assumed that the human has left, and the track is deleted. In this application, while the robot is moving and navigating the human there may be several sharp turns and corners. So due to a lack of space behind the robot, humans may not exactly stand behind the robot and may be out of the field of view of the camera. Here the human cannot be detected by the robot. So, the time of invisibility is not a suitable metric. In this chapter, the possible intents of the robot and the person are detected. If there is enough evidence that the person can follow the robot and the person is not doing the same, only then the track is deleted. This saves a lot of false positives that can have a strong advantage from an application perspective.

If the human leaves the group and re-joins the group, then the tracker is also deleted and re-initialised. In the initial phase, the robot is assigned the initial position of the human, and when the tracker is re-initialised, the robot detects the human in the 2D image plane and then converts this 2D image plane coordinate into a 3D real-world coordinate, following which a continuous tracking is performed. In this approach, only a monocular camera is used to track the human and to determine the position of the human in the 3D real-world. The 3D position can be estimated with knowledge of the 2D position in the image with the assumption that the height of the person is known.

It may be curious to know why the thesis uses only a single monocular camera, when several cameras can be fitted onto the robot. There is one important reason for using a single camera which is to reduce the computation time of the algorithm. Multiple cameras will increase the computation time because lot of signal data will flow into the algorithm which is difficult to handle in a near-real time. In contrast, the benefits are limited. There are only a few cases where the robot needs to make sharp turns leading to a loss of visibility. Many such cases may involve obstacles that will inhibit visibility even if multiple cameras are mounted on the robot. The thesis suggests that it is better to contextually handle the cases, rather than requiring additional cameras that will increase the overall computation time even in normal cases.

Here the key performance indicator is the tracking error on a dataset with the same challenges as discussed in section 3.1.2 with the additional challenge of erratic Human Behaviours. An additional key performance indicator is the ability to detect people leaving the group measured using the standard binary classifier metrics (true positive, true negative, false positive, false negative, precision, recall, accuracy, and F-score).

Experiments confirm that the proposed methodology detects people leaving the group more accurately and tracks the people more accurately as compared to several baseline approaches. The differences are particularly large when the robot has a limited field of view of the subjects for a long time within the experiment. The proposed approach more accurately detects people leaving the group and performs a better social tracking when compared to several baseline approaches.

3.1.4 Organization

Section 3.2 summarises the related works carried out in the literature. Sections 3.3 and 3.4 explain the face tracking approach with different heights of the camera and vehicle tracking. The Human Following Robot (HFR) behaviour-based tracking and its algorithm is explained in section 3.5. Section 3.6 describes the multi-behaviour human tracking in a group. Section 3.7 elaborates the multi-behaviour human tracking model and presents its architecture. Experiment evaluation and result analysis of each social tracking approach is discussed in section 3.8. Section 3.9 gives the conclusion of this chapter.

3.2 Related Work

Many tracking algorithms have been developed based on background subtraction using one or several stationary cameras. In these systems, tracking of people or objects is performed which is based on the extracted component that belongs to the foreground. In general, object tracking can be categorised as: Active contour-based tracking, Model based tracking (Lou et al., 2005), Feature based tracking, and Region based tracking. A Kalman filter can also be applied for object tracking, it can predict the behaviour of a moving object categorised by a predefined model (Chen et al., 2018).

3.2.1 Vehicle tracking

In the intelligent transportation system, vehicle tracking is a modern research work in recent years. Since there is a very complex traffic environment, so due to this it is very difficult to track multiple vehicles robustly and accurately. To overcome these challenges two classifier and improved ViBe algorithms (Min et al., 2017) were used because the ViBe algorithm found it difficult to differentiate between multiple moving vehicles. This algorithm removed the object residual shadows and restrained the dynamic noise. This method had the time efficiency advantage of the SVM and the accuracy advantage of the CNN.

Xie et al. (2017) utilised a Physics and manoeuvre-based model with the integration of an Unscented Kalman Filter (UKF) and Monte-Carlo method for the prediction of the vehicle trajectory. Moreover, a vehicle kinematic model and driving behaviour awareness network have been developed. Bencherif and Chouireb (2019) used a fuzzy neural network with the kinematic model and parameter learning for the tracking of a robot trajectory.

Vehicle tracking has an excellent application in the parking scheme. To develop this, a VeTrack (real time vehicle tracking system) was proposed (Gao et al., 2017). A probabilistic algorithm (Shan et al., 2014) was also developed to track multiple vehicles in a wide range area which was based on the vehicle model that includes the surrounding traffic environment along with timing profile, velocity, and acceleration by using the historical data collected.

3.2.2 Face Detection and Tracking

Face recognition includes three main steps: face detection, face recognition, and face extraction. In a video sequence a face is detected by a multiple method such as Viola Jones Face Detection Algorithm (Viola and Jones, 2001), Local Binary Pattern (LBP) (Ahonenet et al., 2004), AdaBoost Algorithm for Face Detection (Meir and Rätsch, 2003). Here face tracking is considered to be an optimisation problem, so to resolve the problem of face tracking a meta-heuristic optimisation algorithm and differential harmony search (DHS) was introduced by Gao et al. (2014). This algorithm was compared with the Mean Shift Filter, Particle Filter, HIS (Fourie et al., 2010) and it was found that a DHS based tracker gives a more accurate result. Silhouette image and edge detection are widely used for human motion tracking. Liu et al. (2016) applied a filtering algorithm with stereo calibration with the Markov model to perform human tracking. Further, Castillo et al. (2017) performed human detection and tracking in a distinct environment.

Shi et al. (2016) performed detection and tracking of faces that have different angles (initially). Distinct faces were matched to the face in frame. Kansal et al. (2012) developed an approach for target tracking of the human using its upper body torso colour feature and face recognition. Aghaei et al. (2016) also presented a multi face tracking approach that generates a tracklets group and takes advantage of tracklets redundancy to handle the unreliable ones.

The human following framework consists of three attributes: human detection, human recognition, and human tracking (Chi et al., 2017). In these circumstances, target recognition is a challenging task for following humans. Yuan et al. (2019) combined principal component analysis with tracking learning detection and applied on the human face (Face-TLD) to achieve an improvement which is called IFace-TLD. The IFace-TLD improved the discrimination ability of the Face-TLD, such as ambiguous facial appearances. Moreover, head pose estimation (Yu et al., 2018) was used for 3D tracking pose estimation. A multitask convolution network (Zhang et al., 2016; Li et al., 2015) was applied for face detection and alignment.

Instead of trying to develop better detectors for the human, the proposed work emphasises on a better prediction of the human trajectory that is especially useful since the person can be within the frame of view for a limited time only. For this, the proposed work model group behaviours where a group of people follows a robot, including an accurate detection of people leaving the group and tracking even in low-field-of-view conditions. The group leaving detection is primarily driven by an understanding of the context of the robot operation.

3.2.3 Human Tracking using Depth Information

Liu et al. (2016) captured and tracked the human motion based on multiple RGBD cameras. The method could also solve the problem of body occlusion. Furthermore, Yu et al. (2019) developed an algorithm to capture very fast human motion based on a single RDB-D camera. The method concatenated pose detection with motion tracking. Moreover, human motion prediction can also be carried out by the modelling of convolutional hierarchical auto-encoder (Li et al., 2019). A very popular method is to detect the legs by using the moving blobs detection which also appears as local-minimal in range data (Lenser and Veloso, 2000; Jensfelt et al., 2000; Wengefeld et al., 2019).

The robots can be applied to different servicing areas if they are reliable and safely follow humans. Sung and Chung (2015) addressed the tracking of humans in a cluttered and dynamic scenario where the tracking algorithm considered the interdependence between the targets. The design module of this tracking was a grouped target which additionally consisted of a leg-pair target. So, a group of two legs as a target was considered as a human. When the human legs are observed from the LRF (Laser Range Finder), then the targets (human legs) are not independent. Specifically, in a crowded and cluttered environment, the human has to be sorted among the measurements which are observed. Here the authors developed HSJPDAF (Hierarchical Sample-Based Joint Probabilistic Data Association Filter) for robust tracking, which was able to predict the target in relation to their legs. HSJPDAF was divided into two levels, low and high. At a low level, the observation model was used to predict the state of targets legs while at the high level, the states of group target ID were determined with the association of the observation model.

3.2.4 Dense Human Tracking and Tracking for Motion Flow

It is hard to detect and track the human motion and their behaviour in real-time as the robots and vehicles co-exist with multiple humans (Bera and Manocha, 2014). The human behaviour is not fixed and it changes, which makes the tracking difficult. Humans always try to change their velocity to avoid obstacles and other moving humans. Here a few tracking approaches have been proposed which were very reliable and worked for non-real-time and offline applications (Bera and Manocha, 2014). In real-time tracking, the trajectory of every moving human is based upon their sub-goal positions and basic interactions with other humans and obstacles. Therefore, an appropriate motion model should be built for accurate crowd tracking which must include these characteristics. Motion models which do not account for the unpredictability of human trajectories in a large, crowded scenario also include several broadly applied motion models based on constant acceleration and constant velocity model break-down to utilise these human behaviours (Bimbo and Dini, 2011). The proposed approach is a step in the same direction of socially modelling the person, applied on a camera on the moving base (robot), which makes the problem harder. The proposal is for a specific application where the human follows the robot, which is a largely untouched problem.

A MotionFlow model (Jang et al., 2015) was introduced based on a visual analytic system and a different pattern analysis of human motion and further customisation of a robot appearance (Guo et al., 2017). Human pose clustering over multiple motion sequences and its encoding has been applied for tracking data to visualise the motion flow (Jang et al., 2015). Here motion pattern was also applicable to predict the split, merge, and lock situation of tracking data.

3.2.5 Trajectory prediction

Many recent works use deep learning models to predict the trajectory of several human agents. Sun et al. (2018) developed a 3D network model for the trajectory prediction of the human pose using LSTM. In an unconstrained environment, a CNN model (Kumar and Sukavanam, 2020) was proposed for human detection and tracking. Additionally, memory augmented network (Marchetti et al., 2020) has been modelled and utilised for multiple trajectory predictions on the road and pavement with the injection of CNN encoder. The problem with these methods is the need for a large quantity of noise-free training data, which is not available as per the proposed modelling. This thesis assumes that the human will not always be visible, the human behaviour will be highly erratic, and the captured image of the humans will also be highly noisy.

3.2.6 Motion Planning

The motion model in the proposed approach is closely related to the problem of path planning, where the motion of a moving person is predicted. A constraint in the model motion unlike planning is availability of only a partial knowledge of the world that affects the motion of the person and hence most research is not applicable. In the distance aware dynamic roadmap approach (Knobloch et al., 2018) the collision free path planning was implemented by using the distance from the obstacle to determine the trajectory of the robot. Another approach called the constriction decomposition method (Brown and Waslander, 2016) was derived and tested on complex and indoor environment. Variants of an Artificial Potential Field (Orozco-Rosas et al., 2019) and coordination strategy (Kala, 2018) are used for motion planning.

Here the social robot guides humans in a cluttered and dense pedestrian environment. So, the robot can freeze (Trautman et al., 2015) if the path is not safe and the probability of a collision is high. Kim and Pineau (2016) proposed for socially adaptive path planning in a non-static environment which included the inverse reinforcement learning, feature extraction, and path planning modules. Furthermore, Kretzschmar et al. (2016) developed a new approach that enabled a social robot to learn a model of navigation behaviour of the pedestrians and this approach used the Hamiltonian Markov chain Monte Carlo sampling. In more advancement and achievement of socially-aware navigation, a new algorithm was proposed called the SocioSense which was based on the social constraints (Bera et al., 2017).

Roozegar et al. (2016) proposed a mechanism for path planning which included the reinforcement learning and path planning modules for a mobile robot. Pereira et al. (2019) proposed a real-time 3D trajectory planner in the context of an autonomous vehicle for leader following. Here the planner was based on the basic concept of trailer a body whose hinge point was attached to the leader and every follower was assigned to a different point of the trailer. Admittance control was utilised to implement robot guidance (concept of "walk-through programming") in (Ferraguti et al., 2017; Landi et al., 2016). Moreover, a new motion planning algorithm (Park et al., 2019) was proposed for human-motion prediction. This algorithm was also applicable for determining the smooth and collision-free path for the robot. When multiple robots move and behave like a swarm group, then they can follow a leader robot in a straight line and sort themselves to stand in a line according to alpha-numeric name (identity) as demonstrated in (Kumar et al., 2020). Jeddisaravi et al. (2016) developed a framework for robotic exploration in a given static environment. In a real-time scenario, pedestrian model-based reactive planning was performed for assistive robots (Bevilacqua et al., 2018).

The artificial potential field method is widely employed for the social robot navigation. Pozna et al. (2009) proposed a mathematical modelling and construction of the potential field function for the obstacle avoidance and dynamic modelling of the car. In the context of social robot navigation, the group split and again merge prediction by the 3D convolution neural network was demonstrated by Wang and Steinfeld (2020). Li et al. (2008) analysed several behaviours of the social robot to control the robot.

3.2.7 Other Social Robot Applications

Haidegger et al. (2011) modelled a medical application onto the serving robot that can assist in surgery. The authors also produced reliable and fast control signal which can deal with different changing environmental conditions. The robot can also be employed in medical operation and surgeries. The movement of the robot is an essential task for carrying out a successful job. It is also called as Remote Centre of Motion (RCM) (Dahroug et al., 2017). The authors used a combination of RCM and path following controller. The surgical robot navigates inside the patient body via incision point and performs the surgery. In the development of swarm robotics (Tang et al., 2019), particle swarm optimisation and a fruit fly optimisation algorithm are also used.

Fiorini et al. (2020) performed emotion classification using the ECG, GSR, and BA sensors. Moreover, a hand gesture segmentation approach (Thabet et al., 2018) has been developed for feature fusion in a complex background. Beyond this fuzzy logic (Khaksar et al., 2014), text detection and tracking (Tian et al., 2017), can also be applied for navigation and notification.

3.3 Face Tracking

The first aspect of the work is to track faces in indoor environments. A difference from the standard literature is that only the faces are tracked and not the complete human body. This is because in highly congested situations representing interesting behaviours, while most of the human body is occluded, the face of the person is reasonably clear. Further, this thesis is only interested in the trajectories of the humans when walking and no other dynamic obstacles. Full body motion trackers can seldom track other dynamic obstacles like moving chairs, people moving while rolling on chairs, etc. For the specific application a face-based tracking is the most robust technique of work. Similarly, the specific work can incorporate multi-face tracking, unlike many other popular approaches that are not scalable to multiple humans with part occlusions. In the video sequence multiple faces are detected and tracked using the Kanade-Lukas-Tomasi (KLT) algorithm. The procedure for face tracking is given in the following steps. The steps are repeated for every new frame in the video.

- **I.** Read the next frame.
- **II.** Detect the human face and find the feature descriptors of the face.
- **III.** Assign an identity or box number to every bounding box so that the number of faces can be predicted in a video sequence.
- **IV.** Match the feature descriptors vector with the tracked faces.
- **V.** If the point features match, then track the face and identity of bounding box identity will remain the same.
- **VI.** If another face is found, compute a new feature descriptor and track it.
- **VII.** If the face goes outside the new frame, delete the track.
- **VIII.** Draw a bounding box around the face.

3.4 Vehicle Tracking

The second extreme case of tracking is encountered in tracking of moving vehicles. In conception it may appear that the same tracking methodology as indoor person tracking can be used in this case. However, this is not true since the environment noise is much increased for outdoor scenarios as compared to indoor scenarios, the resolution of operation is much poorer in outdoor scenarios. Further, there is no distinct feature such as face in a traffic scenario; each vehicle in a traffic scenario can vary largely by size and shape. This is especially true in the Indian context where vehicles can vary from a bicycle to a bus. Hence a new technique is designed to combat these limitations. Correspondingly, the method proposed here is not ideal for tracking of faces as the method does not benefit from the generic heuristics of a well-placed indoor camera and the shape of the human face.

The proposed method detects moving vehicles in a video sequence and also associates detection correspondence to the same object over the time. Here a background subtraction method is applied in which the first frame considered as the background frame and the remaining frame as a foreground. The Kalman filter is used to estimate the motion of every track and find the track location in every frame. If a moving person or object is not available after some frames, then the old track is deleted. The object tracking is summarised in the following steps that happen in every time iteration:

I. Read the next video frame.
II. Consider the first frame as a background frame and use frame difference to compute the foreground.
III. Find the location of the object in the foreground using blob analysis.
IV. Update the Kalman filter using the motion and observation models.
V. Draw a bounding box around the moving object with the identity of the bounding box.
VI. If the scope of an object does not exist, then delete the track and the bounding box.
VII. If a new object is found, then add the new track and draw a bounding box around the object and associate the box with an identity.

3.5 Human Following Robot (HFR) behaviour and its algorithm

The basic architecture of the particle filter-based tracking model is summarised in Figure 3.1. Here a new method has applied to track the human in a 3D environment. A social particle filter motion model is proposed to track the human following the robot. Here the robot starts its journey from the initial state (origin (0, 0)), and the human follows the robot while maintaining a certain social distance. The interesting aspect is when either the person or the robot has to make a sharp turn in the workspace, wherein the ideal following behaviour is no longer applicable and the person further gets outside the field of view of the camera. Only the initial position of the human is known by the particle filter, which tracks the human and determines its position in 3D environment using a single rear-facing monocular camera.

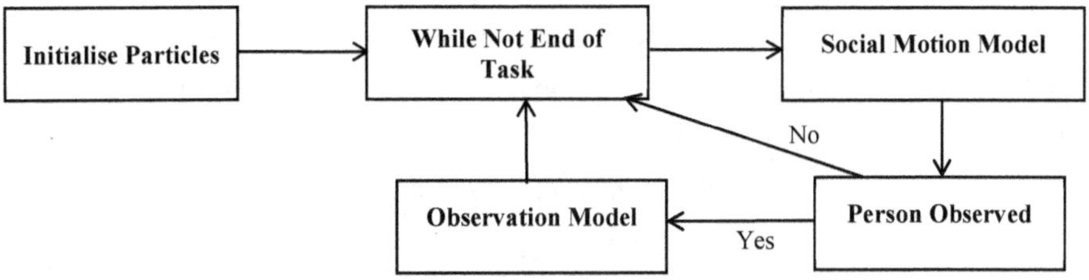

Figure 3.1: Particle filter model

3.5.1 Motion Model of the Particle filter

The first task in the design of the particle filter is to make a motion model. The typical approaches in the literature assume a randomised motion model in which the uncertainties sharply increase in the absence of observation, which is not suitable for the current application with limited sensing. Furthermore, the randomised motion model may not accommodate sharp turns by the humans. The humans are less kinematically constrained and can make sharp turns, which may be needed due to a densely packed workspace or a sudden turn by the robot. The proposed model assumes that the robot position in 3D is known at every instance of time, which is obtained from the localisation module of the robot. While computing the human estimated position the proposed model also considers the robot position.

It is assumed that the person is following the robot, which is a very useful information in order to predict the motion of the human. The human is assumed to follow a strategy to maintain a social distance from the robot. The humans tend to always keep a social distance between themselves while navigating. The social distance is complex and depends upon the affinity between the two people and the group to which they belong. Fortunately, the robot is not considered as socially amicable and not a part of any social group involving humans. Hence let the desired separation between the human and robot be given by d_{soc}, which includes the radius of the person and robot. The notations are shown in Figure 3.2.

Some experiments were carried out with human subjects to get the social distance d_{soc}. The distance, unfortunately, depends upon the particular subject and situation. The distance is hence sampled from a uniform distribution. For modelling the social distance is however taken as a constant (represented by the mean alone) and uncertainties are added to the other noise terms.

Let $q_t^{(i)}$ be the pose of the person as inferred by particle i at any time t. Let the pose of the robot be given by r_t. The person immediate speed is set so as to attain a social distance d_{soc} eventually, which is modelled by using Equations (3.1-3.3) using the principles of a proportionate controller having a proportionality constant K_P.

$$v_t^{(i)} = K_P \max\left(d_t^{(i)} - d_{soc}, 0\right) + \epsilon_v \tag{3.1}$$

$$d_t^{(i)} = \sqrt{\left(q_{t,X}^{(i)} - r_{t,X}\right)^2 + \left(q_{t,Y}^{(i)} - r_{t,Y}\right)^2} \tag{3.2}$$

$$\epsilon_v \sim U[-\epsilon_{vmax}, \epsilon_{vmax}] \tag{3.3}$$

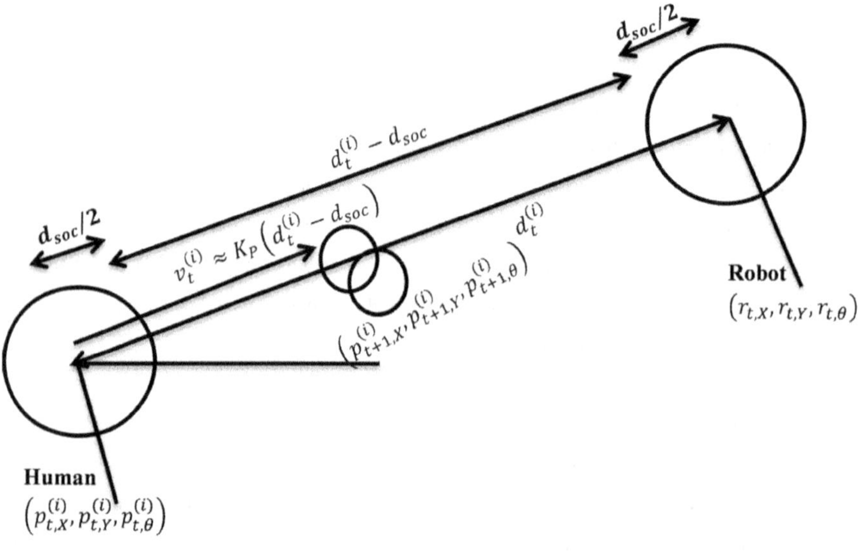

Figure 3.2: Particle filter behavioural module and its geometry

Here d_{soc} is the social distance between the person and the robot including the radii of the two entities. $d_t^{(i)}$ is the distance between robot and the human (particle). A restriction applied in the selection of speed is that the person does not go backward, which is another social etiquette when people follow each other. Temporarily the distance may be less than required, but the robot eventually moves ahead increasing the distance. ϵ_{motion} is a noise allowing for the randomised nature of d_{soc}, the change of behaviour of the person, momentary distraction, accounting for person personality specific characteristics, presence of obstacles, etc. The noise is sampled from a uniform distribution $[-\epsilon_{vmax}, \epsilon_{vmax}]$.

The speed of the person is subjected to a threshold $v_{\max}$, further constraining it not to be negative, given by Equation (3.4)

$$v_{t,lim}^{(i)} = \max\left(\min\left(v_t^{(i)}, v_{max}\right), 0\right) \quad (3.4)$$

A heuristic is that the person cannot move faster than the maximum speed of the robot, while the maximum speed of the robot is available from the robot documentation. This helps in setting the value of $v_{\max}$.

The person is expected to turn so as to face the robot. Unlike robots, the humans do not have non-holonomic constraints and are able to make sharp turns. The person hence is assumed to have immediately oriented in the direction of the robot, with an infinite angular speed, and thus the orientation is given by Equations (3.5-3.6).

$$q_{t+1,\theta}^{(i)} = atan2\left(r_{t,Y} - q_{t,Y}^{(i)}, r_{t,X} - q_{t,X}^{(i)}\right) + \epsilon_\theta \quad (3.5)$$

$$\epsilon_\theta \sim U[-\epsilon_{\theta max}, \epsilon_{\theta \max}] \quad (3.6)$$

Here ϵ_θ is a noise sampled from a uniform distribution $[-\epsilon_{\theta max}, \epsilon_{\theta \max}]$. The noise accounts for factors including smoothness preference of the human, specific behaviour characteristics, presence of obstacles, etc.

The predicted position of the particle is thus approximately given by Equations (3.7-3.8);

$$p_{t+1}^{(i)} = f_{soc}\left(p_t^{(i)}, r_t\right) \quad (3.7)$$

$$\begin{bmatrix} p_{t+1,X}^{(i)} \\ p_{t+1,Y}^{(i)} \\ p_{t+1,\theta}^{(i)} \end{bmatrix} = f_{soc}\left(\begin{bmatrix} p_{t,X}^{(i)} \\ p_{t,Y}^{(i)} \\ p_{t,\theta}^{(i)} \end{bmatrix}, \begin{bmatrix} r_{t,X} \\ r_{t,Y} \\ r_{t,\theta} \end{bmatrix}\right) = \begin{bmatrix} p_{t,X}^{(i)} \\ p_{t,Y}^{(i)} \\ 0 \end{bmatrix} + \begin{bmatrix} v_t^{(i)} \cos p_{t+1,\theta}^{(i)} \\ v_t^{(i)} \cos p_{t+1,\theta}^{(i)} \\ atan2\left(r_{t,Y} - q_{t,Y}^{(i)}, r_{t,X} - q_{t,X}^{(i)}\right) + \epsilon_\theta \end{bmatrix} \quad (3.8)$$

It is also good to discuss the randomised motion model, which simply adds noise to produce the state of the person at any instant of time, given by Equations (3.9-3.12);

$$p_{t+1}^{(i)} = f_{rand}\left(p_t^{(i)}\right) \quad (3.9)$$

$$\begin{bmatrix} p_{t+1,X}^{(i)} \\ p_{t+1,Y}^{(i)} \\ p_{t+1,\theta}^{(i)} \end{bmatrix} = f_{rand}\left(\begin{bmatrix} p_{t,X}^{(i)} \\ p_{t,Y}^{(i)} \\ p_{t,\theta}^{(i)} \end{bmatrix}\right) = \begin{bmatrix} p_{t,X}^{(i)} \\ p_{t,Y}^{(i)} \\ 0 \end{bmatrix} + \begin{bmatrix} v_t^{(i)} \cos p_{t+1,\theta}^{(i)} \\ v_t^{(i)} \cos p_{t+1,\theta}^{(i)} \\ \epsilon_\theta \end{bmatrix} \quad (3.10)$$

$$v_t^{(i)} \sim U[-v_s, v_s] \quad (3.11)$$

$$\epsilon_\theta \sim U[-\theta_s, \theta_s] \tag{3.12}$$

Here v_s and θ_s is the maximum noise related to speed and angle respectively.

A purely social behaviour may not be able to deal with non-anticipated changes such as the person circumventing an obstacle, the person deliberately showing non-cooperation, etc. Hence a fusion of the social and random motion model is also used, given by Equations (3.13-3.14);

$$p_{t+1}^{(i)} = w_1 f_{soc}\left(p_t^{(i)}, r_t\right) + (1 - w_1) f_{rand}\left(p_t^{(i)}\right) \tag{3.13}$$

$$w_1 \sim U[0,1] \tag{3.14}$$

Here w_1 controls the contribution of the social and randomised terms.

3.5.2 Observation Model of the Particle Filter

The motion model predicts the human position in the 3D real world coordinates and this 3D point is projected into the 2D image plane. The proposed approach is applied on a monocular camera which is calibrated so that lens distortion and intrinsic parameter are available as a priori information. Hence a standard calibration matrix (C) is given by Equation (3.15) which is used in the proposed camera model,

$$C = \begin{pmatrix} f_x & p_s & ¢_x & 0 \\ 0 & f_y & ¢_y & 0 \\ 0 & 0 & 1 & 0 \end{pmatrix} \tag{3.15}$$

Here $(¢_x, ¢_y)$ is the principal point offset and f_x and f_y represents the focal length while p_s is axis skew (shear distortion in the projected image).

The robot coordinate frame is fixed to the centre of the differential wheel drive of the robot. The localisation module available with the robot gives the transformation between the robot and the world coordinate axis system at any point of time. The camera is facing the rear and thus has a rotation of π along the Z-axis, and is at a height h_c. This gives the transformation between the camera and the world coordinate axis as expressed as Equation (3.16);

$$T_{camera'}^{W} = \begin{pmatrix} cos(R_t.\theta + \pi) & -sin(R_t.\theta + \pi) & 0 & R_t.X \\ sin(R_t.\theta + \pi) & cos(R_t.\theta + \pi) & 0 & R_t.Y \\ 0 & 0 & 1 & -h_c \\ 0 & 0 & 0 & 1 \end{pmatrix} \tag{3.16}$$

Here $R_t.\theta$ is the robot's orientation.

As per convention, the Z-axis of the camera goes out of the camera and faces the human. Correction to the same conventions is expressed by the transformation given as Equation (3.17);

$$T_{camera}^{camera'} = \begin{bmatrix} 0 & 0 & 1 & 0 \\ 1 & 0 & 0 & 0 \\ 0 & 1 & 0 & 0 \\ 0 & 0 & 0 & 1 \end{bmatrix}^T \tag{3.17}$$

The human prospectively at position $h_{t,p}^{(i)}$ should hence be visualised at a position $T_{camera'}^{W} h_{t,p}^{(i)}$ in the camera coordinate axis system which is further changed to $T_{camera}^{camera'} T_{camera'}^{W} h_{t,p}^{(i)}$ as per the camera coordinate frame. Here projection onto an image takes place by passing through the camera calibration matrix and producing the image point as given in the Equations (3.18-3.20);

$$\begin{bmatrix} u' \\ v' \\ s' \end{bmatrix} = C T_{camera}^{camera'} T_{camera'}^{W} h_{t,p}^{(i)} \tag{3.18}$$

$$u_{t,p}^{(i)} = \frac{u'}{s'} \tag{3.19}$$

$$v_{t,p}^{(i)} = H - \frac{v'}{s'} \tag{3.20}$$

Here H is the height of the image and the subtraction is used to denote the change that the object detection uses the top-left as the origin, while the camera uses the bottom left as the origin. The coordinate systems used in the computation are represented by Figure 3.3.

The overall complete Pseudo code of this tracking and detection module is explained briefly as Algorithm (3.1).

Algorithm 3.1: Particle filter module

1. Determine initial position of the person $(person_x, person_y, person_\theta)^T$
2. $t \leftarrow 0$
3. For all particles *i*
   - **A.** $p_t^{(i)} = (person_x, person_y, person_\theta)^T$
   - **B.** $w_t^{(i)} = 1/n$
4. Initialise the camera
5. While (human is following the robot)
   - **A.** For every particle *i*, Calculate $p_{t+1}^{(i)}$ using (3.7), (3.9) or (3.13)
   - **B.** $z_{t+1} \leftarrow faceDetector()$
   - **C.** $If(z_{t+1} \neq null)$
     - i. *For every particle i*
       - **a.** Calculate $(u_{t+1}^{(i)}, v_{t+1}^{(i)})$ from (3.22) and (3.23)
       - **b.** Calculate $w_{t+1}^{(i)}$ from (3.21)
       - **c.** resample particles
   - **D.** *Calculate estimated position* $\bar{q_t} = \sum_{i=1}^{n} w_t^{(i)} q_t^{(i)}$
   - **E.** $t \leftarrow t + 1$

Here face detection method is applied to detect the human face. The observation model is applied here to assign weights to the particles. Let $z_{t+1}=(u^{obs}_{t+1}\ v^{obs}_{t+1})$ be the observed centre of the bounding box of the face in the 2D image plane. The updated weight of the particle i is given by Equation (3.21);

$$w_{t+1}^{(i)} = p\left(z_{t+1}|q_{t+1}^{(i)}\right) = \frac{\exp\left(-\frac{\left(u_{t+1}^{obs}-u_{t+1}^{(i)}\right)^2+\left(v_{t+1}^{obs}-v_{t+1}^{(i)}\right)^2}{2\sigma^2}\right)}{\sum_i \exp\left(-\frac{\left(u_{t+1}^{obs}-u_{t+1}^{(i)}\right)^2+\left(v_{t+1}^{obs}-v_{t+1}^{(i)}\right)^2}{2\sigma^2}\right)} \tag{3.21}$$

Where

$$u_{t+1}^{(i)} = u_{t,p}^{(i)} \tag{3.22}$$

$$v_{t+1}^{(i)} = v_{t,p}^{(i)} \tag{3.23}$$

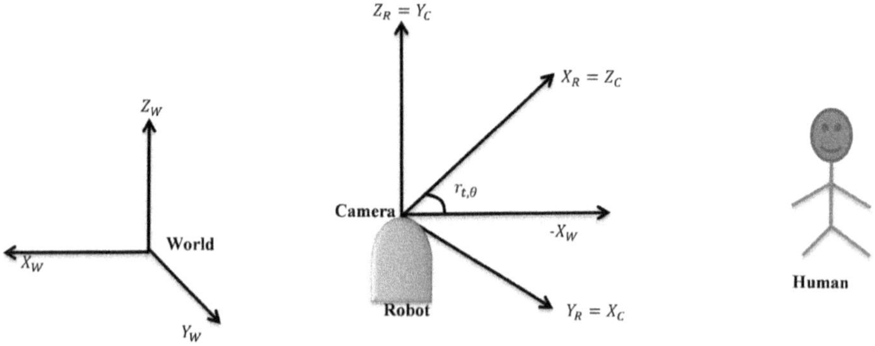

Figure 3.3: 3D coordinate system of human and robot

3.6 Multi-behaviour humans tracking in a group

The work is now extended to multiple people in the group. The major problem associated with multiple people is that they display multiple behaviours. Now even if a person may like to follow the robot, it may not be possible due to other people not giving enough space. The people may have non-ideal behaviours such as temporarily leaving the group or being temporarily distracted. The implementation of the particle filter specifically asks for a motion model to first anticipate the motion of the person hypothesised in a generic sense by the particle and then to move the particle using the same anticipated model. The motion model has a high value for the proposed work since the person will often be outside the visible region and the accuracy of the system largely depends upon the motion model to accurately predict the motion of all persons. The motion model discussed in Sec. 3.6.1 uses the concept of a social potential field that measures how much a person is attracted and repelled by the other people in the vicinity and the robot to predict the direction of motion.

The next important aspect of a Particle Filter is to correct the errors developed in the motion model and is carried out by using an observation model. Every particle mentions a hypothesised position of the person. The observation model projects the hypothesised positions of the persons onto the image taken by the camera, given the location of the robot and hence the location of the camera. If the hypothesis was good, the projected position of the person will match the observed position on the image and vice versa. This is used to re-weight the particles. After re-weighing the particles are re-sampled which increases the particle density in good areas and decreases the particle density in poorer areas.

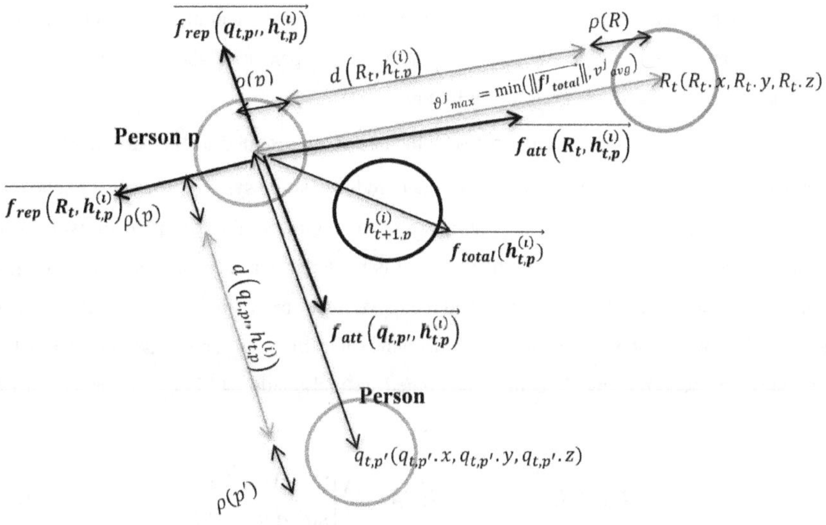

Figure 3.4: Multi-behaviour based computational geometry of 3D-PSFM

This thesis has developed a framework using multi-behaviour 3D particle social force model which consists of human detection, analysis of human behaviour, and different force applied to compute human position in 3D real-world and 2D image plane. The different modules are discussed in the following sub-sections.

3.6.1 Multi-behaviour Motion Model

The purpose of the motion model is to predict the control input $\mathcal{U}_{t,p}^{(i)}$ for particle i and person p at time t, given that the state of the system is being continuously tracked. The person may not be visible for a long duration of time and therefore the onus of correctly tracking the person relies upon the motion model alone. The motion model is made using the social potential field algorithm, where essentially every particle witnesses a social force from different factors that govern the overall motion of the particle. A particle representing a person's position feels an attraction towards the robot that is dominant when the person is far away. The person also feels a repulsion on being too close to the robot. Similarly, the person feels an attraction from the other people in the vicinity on being far away. This behaviour makes the group stay intact. Similarly, the people feel repulsion between each other on being too close, which is also often observed as people like to have some personal space. The particle is moved based on the cumulative effect of all these forces. This model is used to anticipate the state of the person at the next time instant. A representation of the notations is shown in Figure 3.4.

The first force is the attraction between the person represented by the particle $h_{t,p}^{(i)}$ and the robot whose accurate pose R_t is known by the onboard localisation system. An only attraction may make the person come too close to the robot, almost kissing the surface of the robot, which is not practical. Therefore, an additional repulsive force is modelled that becomes dominant when the robot is too close. Together the attraction and repulsive forces between the person and robot make the person follow the robot maintaining a comfortable social distance. The attraction and repulsion force between the robot and human is computed as in Equation (3.24) and Equation (3.25) respectively

$$\overrightarrow{\boldsymbol{f}_{attr}\left(\boldsymbol{R}_t, \boldsymbol{h}_{t,p}^{(i)}\right)} = w_1 d\left(R_t, h_{t,p}^{(i)}\right)\begin{bmatrix}\cos\alpha\left(R_t, h_{t,p}^{(i)}\right)\\ \sin\alpha\left(R_t, h_{t,p}^{(i)}\right)\end{bmatrix} \tag{3.24}$$

$$\overrightarrow{\boldsymbol{f}_{rep}\left(\boldsymbol{R}_t, \boldsymbol{h}_{t,p}^{(i)}\right)} = -w_2 exp\left(\frac{-d^2\left(R_t, h_{t,p}^{(i)}\right)}{2\sigma^2}\right)\begin{bmatrix}\cos\alpha\left(R_t, h_{t,p}^{(i)}\right)\\ \sin\alpha\left(R_t, h_{t,p}^{(i)}\right)\end{bmatrix} \tag{3.25}$$

w_1 and w_2 are constants that denote the relative importance of the two forces. σ is another constant that determines how quickly the repulsive force fades with an increase of the distance. Here the distance between the particle $h_{t,p}^{(i)}$ and robot R is given by Equation (3.26),

$$d\left(R_t, h_{t,p}^{(i)}\right) = \sqrt{(R_t.x - h_{t,p}^{(i)}.x)^2 + (R_t.y - h_{t,p}^{(i)}.y)^2} - (\rho(R) + \rho(p)) \tag{3.26}$$

$\rho(R)$ is the radius of the robot and $\rho(p)$ is the radius of the human p. $\alpha\left(R_t, h_{t,p}^{(i)}\right)$ is orientation between the robot and human, given by Equation (3.27);

$$\alpha\left(R_t, h_{t,p}^{(i)}\right) = atan2\left(R_t.y - h_{t,p}^{(i)}.y, R_t.x - h_{t,p}^{(i)}.x\right) \tag{3.27}$$

For larger groups, it is further observed that every person in the group cannot follow the robot as there is little space and people cannot overlap. Furthermore, people tend to keep a comfortable space between themselves even though they may not be avoiding a collision. Often people stick together as a group due to an attraction between themselves. In such a case a person may not only follow the robot but also wait for members of the group lagging behind. Therefore, there is repulsion and attraction between every two people in the group. The attraction and repulsive force among every human in the group is given by Equations (3.28-3.29);

$$\overrightarrow{\boldsymbol{f}_{attr}\left(\boldsymbol{q}_{t,p'}, \boldsymbol{h}_{t,p}^{(i)}\right)} = w_3 d\left(q_{t,p'}, h_{t,p}^{(i)}\right) \begin{bmatrix} \cos\alpha\left(q_{t,p'}, h_{t,p}^{(i)}\right) \\ \sin\alpha\left(q_{t,p'}, h_{t,p}^{(i)}\right) \end{bmatrix} \tag{3.28}$$

$$\overrightarrow{\boldsymbol{f}_{rep}\left(\boldsymbol{q}_{t,p'}, \boldsymbol{h}_{t,p}^{(i)}\right)} = -w_4 exp\left(\frac{-d^2\left(q_{t,p'}, h_{t,p}^{(i)}\right)}{2\sigma'^2}\right) \begin{bmatrix} \cos\alpha\left(q_{t,p'}, h_{t,p}^{(i)}\right) \\ \sin\alpha\left(q_{t,p'}, h_{t,p}^{(i)}\right) \end{bmatrix} \tag{3.29}$$

w_3 and w_4 are the proportionality constant denoting the importance of the force. σ' determines how quickly the repulsive force fades with an increase of the distance. $d\left(q_{t,p'}, h_{t,p}^{(i)}\right)$ and $\alpha\left(q_{t,p'}, h_{t,p}^{(i)}\right)$ are distance and orientation between the person p' (as tracked by the system) to the human being tracked represented by the particle $h_{t,p}^{(i)}$, given by Equations (3.30-3.31);

$$d\left(q_{t,p'}, h_{t,p}^{(i)}\right) = \sqrt{(q_{t,p'}.x - h_{t,p}^{(i)}.x)^2 + (q_{t,p'}.y - h_{t,p}^{(i)}.y)^2} - (\rho(p) + \rho(p')) \tag{3.30}$$

$$\alpha\left(q_{t,p'}, h_{t,p}^{(i)}\right) = atan2\left(q_{t,p'}.y - h_{t,p}^{(i)}.y, q_{t,p'}.x - h_{t,p}^{(i)}.x\right) \tag{3.31}$$

The above-mentioned forces create a strong bias on the motion of the people. People when following a robot also show behaviour that is unusual and cannot be directly accounted for. As an example switching places, overtaking the robot, overtaking another person, lagging behind to spend more time at an earlier exhibit, etc. To account for all such forces, a random force has been also utilised here (Equation (3.32)),

$$\overrightarrow{\boldsymbol{f}_{rand}(\boldsymbol{h}_{t,p}^{(i)})} = w_5 r \begin{bmatrix} \cos\theta_{rand} \\ \sin\theta_{rand} \end{bmatrix} \tag{3.32}$$

Here $r \sim U[0,1]$ is a random magnitude of the force taken from a uniform distribution between 0 and 1; and $\theta_{rand} \sim U[-\pi, \pi]$ is a random direction of the force taken from a uniform distribution. w_5 is the relative importance of the force.

The attractive and the repulsive forces add up. However, the random force is uncertain. The force is seldom applied, but once applied has a strong effect. Therefore, a weighted fusion of the random force is made with all the other forces, where the weight is taken as $W_t \in [0,1]$. The different particles will have different weights and therefore different particles shall represent different hypotheses, while the one that matches the observation shall eventually have a large weight. The total social force is given as Equation (3.33),

$$\overrightarrow{\boldsymbol{f}_{total}(\boldsymbol{h}_{t,p}^{(i)})} = \boldsymbol{W}_t \left(\overrightarrow{\boldsymbol{f}_{att}(\boldsymbol{R}_t, \boldsymbol{h}_{t,p}^{(i)})} + \overrightarrow{\boldsymbol{f}_{rep}(\boldsymbol{R}_t, \boldsymbol{h}_{t,p}^{(i)})} + \boldsymbol{\Sigma}_{p'} \left(\overrightarrow{\boldsymbol{f}_{att}(\boldsymbol{q}_{t,p''}, \boldsymbol{h}_{t,p}^{(i)})} + \overrightarrow{\boldsymbol{f}_{rep}(\boldsymbol{q}_{t,p''}, \boldsymbol{h}_{t,p}^{(i)})} \right) \right) + (\boldsymbol{1} - \boldsymbol{W}_t) \overrightarrow{\boldsymbol{f}_{rand}} \tag{3.33}$$

When the human is being continuously detected, the weight has a high value favouring the attraction and repulsive forces as per modelling and is thus given by a normal distribution $W_t \sim N(W_{th}, \sigma_{th})$ with a large magnitude W_{th} and a small variance σ_{th}. However, if the person is not being continuously detected, randomness needs to be added to indicate the increasing uncertainty in the system and is carried out by favouring the random forces by setting the weight as a uniform distribution as $W_t \sim U(W_{min}, W_{max})$ with relatively smaller values.

The force values are converted into the desired linear and angular speeds. The linear speed is given by the magnitude of the force, which is subjected to the constraints of the human walking speed, given by Equation (3.34)

$$\vartheta(h_{t,p}^{(i)}) = \min\left(\left\| \overrightarrow{\boldsymbol{f}_{total}(\boldsymbol{h}_{t,p}^{(i)})} \right\|, v_{max} \right) \tag{3.34}$$

The humans are holonomic and do not have kinematic constraints with respect to orientation. Therefore the humans are said to move in the desired angle modelled by the force, given by (Equation (3.35))

$$\theta(h_{t,p}^{(i)}) = atan2\left(\boldsymbol{f}_{total}\left(\boldsymbol{h}_{t,p}^{(i)}\right).\boldsymbol{y}, \boldsymbol{f}_{total}\left(\boldsymbol{h}_{t,p}^{(i)}\right).\boldsymbol{x} \right) \tag{3.35}$$

The anticipated position of the human is (Equation (3.36))

$$h_{t+1,p}^{(i)} = h_{t,p}^{(i)} + \vartheta\left(h_{t,p}^{(i)}\right)\delta_t \begin{bmatrix} \cos\theta(h_{t,p}^{(i)}) \\ \sin\theta(h_{t,p}^{(i)}) \end{bmatrix} \tag{3.36}$$

Here δ_t is the time period between the consecutive iterations and is the inverse of the tracking frequency.

3.6.2 Multi-behaviour Observation Model

The role of the observation model is to weigh or evaluate every particle for its likelihood of representing the human position. The weight is high if the particle agrees with the observation and poor otherwise. The proposed model has a rear facing camera and an observation model is applied whenever the camera detects the human. It is assumed that the robot's pose and the pose of the camera is known. The particle represents the position of the person. The position represented by the person is projected onto the known image plane of the camera. If the particle represents an accurate position of the person, the projected position on the image plane and the actual observation should closely match and vice versa. The error between the projected position of the person on the image and the actual observed position on the image is used to weigh the particle. Here observation is same as the discuss in the above section but the observation function is different.

The robot takes an image from the camera which passes through face detection and recognition module to identify the different people. Suppose the pth person is observed in the image as $z_{t,p}=(u_{t,c(o)},v_{t,c(o)})$, where c is the correspondence function. The weight of the particle is taken proportional to the error in observation given by Equation (3.37)

$$w_{t,p}^{(i)} = \eta \exp\left(\frac{\left(u_{t,c(o)}-u_{t,p}^{(i)}\right)^2+\left(v_{t,c(o)}-v_{t,p}^{(i)}\right)^2}{2\sigma''^2}\right) \tag{3.37}$$

Here η is the normalisation term set such that the sum of all weights is unity.

3.6.3 Detecting people leaving the group

The people in the group may often leave, which needs to be detected and the tracks need to be deleted. In this section, this human behaviour is examined and find out whether a human has left the group or not. The intuition is that a person not visible for a long time to the camera is assumed to have left the group. However, sometimes the invisibility would be because of contextual reasons. If the robot is making a sharp turn, it is evident that the person would not be able to run behind the robot. Similarly, if the person is following the robot, the person would be making a similar sharp turn after the robot. In such cases, the invisibility is contextual and such cases are filtered out from being deleted.

Let *visible*(t) denote whether the human was detected at time t or not. ¬*visible*(t) does not mean that the person has left the group, as the detection algorithm may have temporarily failed. If any human is not available in a group for a long duration of time then the human is said to have left the group at time t, given by Equation (3.38).

$$left(t) = \neg visible(t') \forall t - \Delta \leq t' \leq t \tag{3.38}$$

Here Δ is the threshold of time within which if the person is not seen, the person is said to have left the group. A large value of the parameter reduces false positives; however, the system deletes the track a long time after the person has left the group. A small value of the parameter may state the person to have left the group even if the detector temporarily failed due to illumination changes.

The situations are detected when the robot is making a turn due to either of these reasons. Let the $(G_t.X, G_t.Y)$ be the sub-goal that the robot is currently seeking. One expects the robot to always keep an angle of approximately, $atan2(G_t.Y - R_t.Y, G_t.X - R_t.X)$ to directly seek the goal under normal obstacle-free criteria. Some diversion is possible due to noises. However, if the current orientation of the robot is very different from this value, it means that the robot has made or is making a sharp turn for whatever reason possible.

$$turn(t) = true\ iff \cos(\alpha_e(t)) < ð \tag{3.39}$$

$$\alpha_e = atan2(G_t.Y - R_t.Y, G_t.X - R_t.X) - R_t.\theta \tag{3.40}$$

Here the coordinates of the sub-goal that is being pursued by the robot are denoted by $(G_t.X, G_t.Y)$ while the coordinates of the robot are denoted by $(R_t.X, R_t.Y)$ with the orientation of the robot as $R_t.\theta$. ð is the maximum deviation in the natural navigation of the robot and a deviation more than this amount is said to denote a turn.

The leaving condition is therefore modified as Equation (3.41)

$$left(t) = \neg visible(t') \wedge \neg turn(t') \forall t - \Delta \leq t' \leq t \tag{3.41}$$

It is important to note that after the robot has completed the turn, people will still not be visible as they are now making the turn, following the footsteps of the robot. However, humans are fast in doing so and the cumulative relaxation of Δ gives enough time for humans to re-join the group.

3.6.4 Track re-initialisation

Imagine a human left the group and is not seen by the camera. Naturally, the human could have left for any reason, however, since the human has now re-joined, the tracking must continue. Since the track was deleted when leaving the group, a new track is initialised based on the visible position of the human on the image. As the human re-joins the group, the track is re-initialised to start continuous tracking. Here the initial position of the human is to be determined from the 2D image plane, as the human is detected in the 2D image plane. The particle filter is initialised which needs to determine the position of the person in the 3D real world. An inversion of the observation model gives Equation (3.42);

$$h_{t,p}^{(i)} = (T_{camera'}^{W})^{-1}(T_{camera}^{camera'})^{-1}C^{-1}\begin{bmatrix} u' \\ v' \\ s' \end{bmatrix} \quad (3.42)$$

Knowing the height of the person directly gives the scale factor (s'). Knowing the position ($u_{t,p}^{(i)}, v_{t,p}^{(i)}$) where the person was observed in the image gives Equations (3.43-3.44);

$$u' = s'u_{t,p}^{(i)} \quad (3.43)$$

$$v' = s'\left(H - v_{t,p}^{(i)}\right) \quad (3.44)$$

Plugging these values also gives the X and Y coordinates of the person.

3.7 Multi-behaviour human model and its architecture

This thesis has implemented a multi-behavioural social particle filter for 3D human tracking performed from a mobile robot while the human is also following the robot. Here the proposed model initialises from the initial position of the humans and starts tracking. When any human leaves the group then the track is deleted and as the human re-joins the group, the track is re-initialised, and the tracking continues. The model is described in Figure 3.5. The term "Social force-based motion model" is used to refer to the computations described in Sec. 3.6.1. A particle represents a possible position of the human. The model is used to compute the new position of a particle given the positions of the other persons and the robot. The previous position of the particle is known. The particle experiences attraction and repulsion forces from all other persons and robot. The particle is moved based on the same forces- that represents the new position of the particle.

If the person is detected in the image, the particles are re-weighted and re-sampled. The particles that project the person into the image closer to the observed location of the person are better and are given a larger weight. This increases the particle density near the true position of the human. In the case of an occlusion or a lack of detection for any other reason, no re-weighing is carried out. This means that the particles will keep moving as per the motion model, however, there will be no way to make a correction that will continuously increase the deviation. Once the occlusion is removed, the re-weighing and re-sampling continue to correct the sample density around the true position of the person.

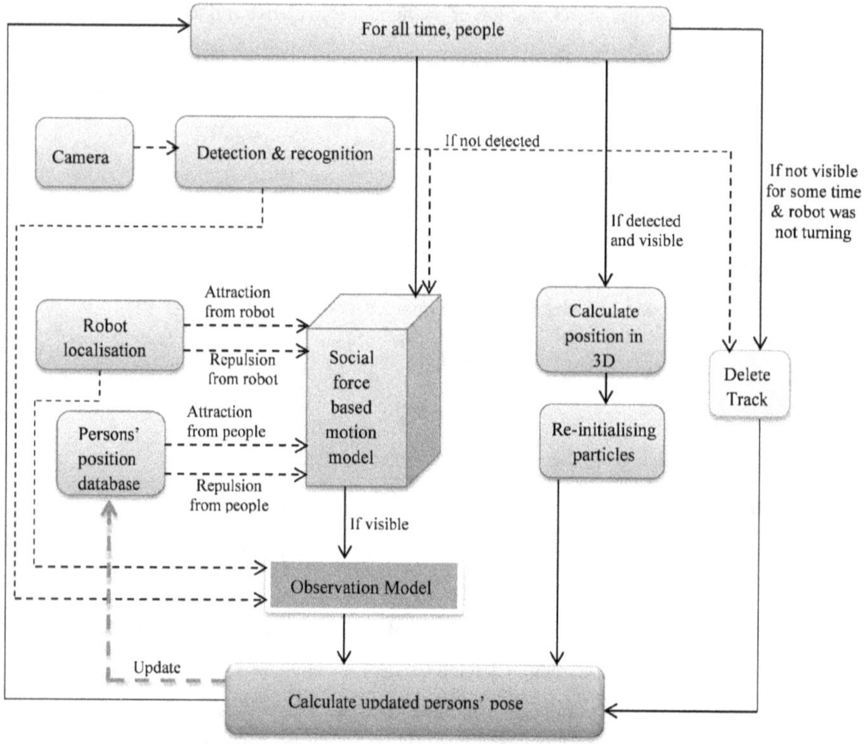

Figure 3.5: Design of Multi-behavioural 3D-PSFM

If a person is far off, he/she is unlikely to be detected by the face recognition library (Schroff et al., 2015; Wen et al., 2016). Furthermore, a person who is far off will not be visible in the camera. In such cases, the robot will not detect the person for a long period of time and the track may get deleted. The robot moves slower than the person. When the person comes closer and becomes visible, a new track will be initialised.

The complete pseudo-code is given as Algorithm 3.2. The algorithm uses a timer to check the time period within which the person p was not visible. The flag *deleted* associated with every person checks if the person is in the group and being tracked, or its track is deleted for leaving the group. $q_{t,p}$ is the tracked position of the person *p* at time *t*, which is used to compute the attraction and repulsion forces between people.

Algorithm 3.2: Tracking

1. For all persons p in group
2. Deleted(p) ← false
3. Start NotVisibleTimer(p)
4. End For
5. For all persons p detected at $(u_{0,p}, v_{0,p})$ and recognised in the first frame
6. Calculate position of person $q_{0,p}$ using (3.42-3.44), or
$$q_{0,p} = \left(T^{W}_{camera'}\right)^{-1}\left(T^{camera'}_{camera}\right)^{-1} C^{-1}\begin{bmatrix} u' \\ v' \\ s' \end{bmatrix}, u' = s'u_{0,p}, and\ v' = s'(H - v_{0,p})$$
7. Initialise particles $h^{(i)}_{t,p} = q_{0,p} + noise, w^{(i)}_{t,p} = 1/n$, where *n* is the number of Particles
8. Reset NotVisibleTimer(p)
9. End For
10. While (tracking)
11. For all people p being tracked
12. Get visible(p), $u_{t,p}, v_{t,p}$ (whether p is visible, and if yes the position in the image) from the face detection and recognition library
13. If Deleted(p) ∧ Visible(p)
14. Calculate position of person $q_{t,p}$ using (3.42-3.44), or
$$q_{t,p} = \left(T^{W}_{camera'}\right)^{-1}\left(T^{camera'}_{camera}\right)^{-1} C^{-1}\begin{bmatrix} u' \\ v' \\ s' \end{bmatrix}, u' = s'u_{t,p}\ and\ v' = s'(H - v_{t,p})$$
15. Initialise particles $h^{(i)}_{t,p} = q_{t,p} + noise, w^{(i)}_{t,p} = 1/n$, where *n* is the number of particles
16. Reset NotVisibleTimer(p)
17. Deleted(p) ← false
18. End If
19. If ¬Deleted(p)
20. For all particles $h^{(i)}_{t,p}$,
21. Calculate $\overrightarrow{\boldsymbol{f}_{attr}\left(\boldsymbol{R}_t, \boldsymbol{h}^{(i)}_{t,p}\right)}$ and $\overrightarrow{\boldsymbol{f}_{rep}\left(\boldsymbol{R}_t, \boldsymbol{h}^{(i)}_{t,p}\right)}$ using (3.24-3.27)
22. Calculate $\overrightarrow{\boldsymbol{f}_{attr}\left(\boldsymbol{q}_{t,p'}, \boldsymbol{h}^{(i)}_{t,p}\right)}$ and $\overrightarrow{\boldsymbol{f}_{rep}\left(\boldsymbol{q}_{t,p'}, \boldsymbol{h}^{(i)}_{t,p}\right)}$ using (3.28-3.31)
23. Calculate $\overrightarrow{\boldsymbol{f}_{rand}(\boldsymbol{h}^{(i)}_{t,p})}$ using (3.32)
24. Calculate $\overrightarrow{\boldsymbol{f}_{total}(\boldsymbol{h}^{(i)}_{t,p})}$ using (3.33)
25. Calculate $h^{(i)}_{t+1,p}$ using (3.34-3.36)
26. End For

```
27.             End If
28.             If ¬visible(p), Update NotVisibleTimer(p)
29.             Calculate Turn(t) using turn(t) = true iff cos(α_e(t)) < ð and α_e =
                atan2(G_t.Y − R_t.Y, G_t.X − R_t.X) − R_t.θ
30.             If Turn(t) calculated using (3.39-3.40), Reset NotVisibleTimer(p)
31.             If NotVisibleTimer(p)>Δ, Deleted(p) ← true (algorithmic
                implementation of (3.41))
32.             If visible(p)
33.                 Calculate u_{t,p}^{(i)} and v_{t,p}^{(i)} using (3.15-3.20)
34.                 Calculate w_{t+1,p}^{(i)} using (3.37)
35.                 Resample h_{t+1,p}^{(i)}, w_{t+1,p}^{(i)}
36.             End If
37.             If ¬deleted(p), q_{t,p} = Σ_{i=1}^{n} w_{t,p}^{(i)} h_{t,p}^{(i)}
38.             Else q_{t,p} = null
39.             End If
40.         End For
41. End While
```

Algorithm 3.2 demonstrates the entire procedure of the 3D-PSF model for multi-humans, trajectory prediction and tracking in a real-time. Initially, every human is detected by their faces in the 2D image plane and the proposed model converts the initial position of every human in the 3D real-world coordinates in lines 5-7. Lines 1-3 start timers useful for the deletion function, however, line 8 resets the timers for the persons visible initially. If a deleted person is further detected in the 2D image plane, then their 3D position is computed from the inverse camera transformation method in lines 13-17. This 3D position is used to re-initialise the tracker for again performing tracking and prediction after deletion of the track. If a person is not deleted, lines 20-26 apply the motion model by using the social force method. Lines 28-31 handle the condition when a person should be deleted because of a lack of visibility. Lines 29-30 handle the case when the robot is making a turn, while line 31 implements the deletion logic. As persons are detected in the subsequent frame then their position in the 2D image plane is determined using 3D real world to 2D image plane transformation (conversion) using lines 31-35. Line 34 uses the visibility information to update the weights while line 35 resamples using the updated weights. Lines 37-40 compute the most probable person location based on the particle position.

3.8 Experiment evaluation and Result analysis of Social Tracking

3.8.1 Tracking using a Stationary Camera

First, the results of tracking of vehicles are presented. A background subtraction algorithm is applied to track the objects and for updating the track of the Kalman filter. An object has its own identity and bounding box. Due to space constrains, only one of the test cases is discussed that represents a sufficient number of moving vehicles along with some static pedestrians. The results are shown in Figure 3.6 (a).

(a) (b)

Figure 3.6: (a) Tracked person (b) Tracked person in cluttered environment

Here, multiple vehicles are detected on the road and some occluded objects are also tracked as shown in the bottom right detection in Figure 3.6 (b). This algorithm did not detect the non-moving pedestrians as shown in Figure 3.6(a) and Figure 3.6(b). They are static persons who were standing for the whole time in the complete video sequence.

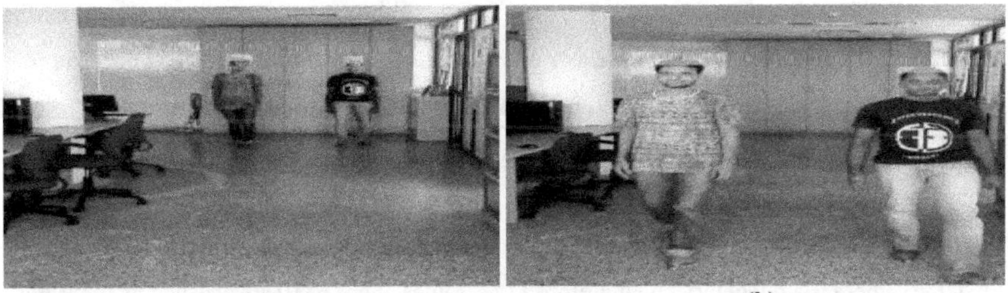

(a) (b)

Figure 3.7: Tracked face of moving two people

The Kanade-Lucas-Tomasi (KLT) algorithm is applied for the face detection on a recorded dataset. Again, only two videos are discussed that contain a small and large number of people who are moving. In the first video two people are moving. The height of the camera from the floor is 1.2 metres, and the distance between person and the camera is 5.4 metres. The detected face is shown in Figure 3.7(a). After some time, the distance between the person and the camera becomes 2.1 metres and the face is detected continuously as shown in Figure 3.7(b).

The same algorithm is also applied to another dataset in which four persons are moving and all detected faces are shown in Figure 3.8. In this dataset, a camera was mounted on the head of the Pioneer robot.

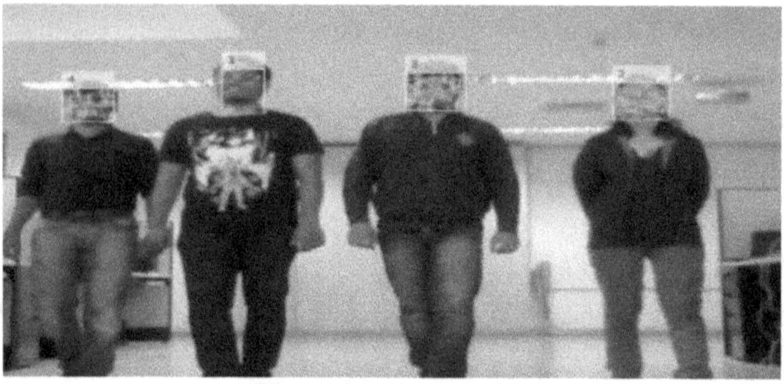

Figure 3.8: Tracked face

3.8.2 Tracking using Human Following Robot (HFR) behaviour

These experimentations are carried out on an Amigobot differential wheel drive robot. The robot localises primarily using high precision wheel encoders. A web camera was mounted on the robot, looking backward. The camera was elevated to nearly human height, which significantly improves the face detection accuracy. The camera was calibrated to get the intrinsic and extrinsic parameters. The initial position of the robot was taken as the origin of the world coordinate frame. The initial setup is shown in Figure 3.9. In this experiment, first the human position was fixed, the robot was moved on the XY axis, and also rotation was applied on the robot using teleoperation using the ROS framework. The person followed the robot. The performance of the algorithm is benchmarked using the key performance indicators considering the specific database challenges. First, the camera is a budgetary and therefore 1128 frames out of 6319 are affected by environment noise. Environment noise refers to the inability of the face detection algorithm to detect the human attributed to lightning condition of the workspace, and shadow problem due to sun light. In the entire workspace the robot also takes several rotations to achieve its goal and therefore the visibility of the human gets lost 11 times for 48 seconds due to the restricted field of view of the camera. The humans move as per the discussed social norms which is already in cooperated in the database for following the robot.

3.8.2.1 Comparisons

The 3D monocular tracking system is tested on different scenarios. Here this thesis discusses three major different scenarios where the proposed method is tested. All testing is carried out at the Centre of Intelligent Robotics of the institute to ensure the validity and correctness of the proposed method. In all scenarios, the human starts walking and follows the robot. In the first scenario (scenario ID I), the robot moved straight ahead for 6 metres and then gradually took an 90° anti-clockwise turn and moved 0.5 metres and then again took an 90° anti-clockwise turn, and thereafter moved straight forward. When the robot covered a 6 metres distance it again took an 60° anti-clockwise turn and stopped. The challenge in this scenario is a repeated frequent turn. First the robot turned and thus the human was out of view, and thereafter the person took time to turn, after a brief period the robot already had to take a second turn. This gave a very brief period for the tracker to re-converge the particles.

(a) (b)

Figure 3.9: AmigoBot robot and its experimental setup with landmark

In the second scenario (scenario ID II), the robot moved straight ahead for 6 metres followed by a gradual 90° anti-clockwise turn and then covered about 3 metres, and again gradually rotated 90° clockwise and covered a 5 metres distance before stopping. The challenge in this scenario is the steep nature of the turn. The robot took too long to fully complete nearly a 180° turn within which there is barely any visibility of the human. In the next scenario (scenario ID III) the robot also covered a major distance with different rotations, specifically took a 360° clockwise rotation. This increases the complexity because the person visibility is lost while the robot is rotating. In this case specifically, the turn is too steep and sudden. Even after the robot completes the turn, the person takes time to turn and come behind the robot.

Two metrics (key performance indicators) are used to assess the performance of the algorithm. The first metric is the trace distance, calculated as the distance between every point in the tracked trajectory with the closest point to the ground truth. The second metric is the inverted trace distance, calculated as the distance between every point in the ground truth to the closest point in the tracked trajectory. The metrics are given by Equations (3.45-3.46); here *d*() is the distance function, $\tau_{tracked}$ is the tracked path and τ_{ground} is the ground truth path.

$$d_{trace} = \frac{\sum_{s} \min_{s'} d\left(\tau_{tracked}(s), \tau_{ground}(s')\right)}{length(\tau_{tracked})} \quad (3.45)$$

$$\overline{d_{trace}} = \frac{\sum_{s'} \min_{s} d\left(\tau_{tracked}(s), \tau_{ground}(s')\right)}{length(\tau_{ground})} \quad (3.46)$$

Table 3.1: Error comparison between proposed method to another method and baseline

Scenarios →	**Scenario ID: I**		**Scenario ID: II**		**Scenario ID: III**	
Name of algorithm ↓	d_{trace} (mm)	$\overline{d_{trace}}$ (mm)	d_{trace} (mm)	$\overline{d_{trace}}$ (mm)	d_{trace} (mm)	$\overline{d_{trace}}$ (mm)
Social Motion	**195.00**	**102.22**	**415.82**	**81.27**	**519.70**	**171.97**
Random motion	1328.70	510.90	2043.51	380.49	2015.41	458.64
0.25-Biased fusion	1173.83	336.43	1863.80	338.86	1901.00	493.14
0.5-Biased fusion	1086.58	377.32	1581.74	282.96	1733.98	460.90
0.75-Biased fusion	960.92	415.61	1411.50	269.51	1567.89	409.51
Fusion	787.95	323.01	1181.03	193.52	1392.81	373.12
Tracking in image	1142.06	298.68	5299.88	194.45	1905.70	307.07

The comparison is made with multiple variants of the proposed algorithm and baselines. The first method is the social particle filter whose motion model is the human following the robot model. The second is the baseline, which is a randomised behaviour with random motion. The third algorithm is the fusion of the social and random motion. The contribution of the social algorithm is made up to a factor of p to get an algorithm parameter, called as the p-biased fusion. The fusion with random weights to social and random components is also used. The last algorithm is a baseline that tracks the humans on the image plane and transforms the results into 3D by estimating the depth of the person. The estimation of depth is possible since the height of the person is assumed to be known, which is enough to calculate the scaling factor.

Table 3.1 gives the error between the particle to robot and robot to particle on distinct algorithms (all the measurements in the mm). The proposed social particle filter approach has the smallest error as compared to all other approaches and baselines. This includes both the error metrics. The worst performance was always from the baseline approach of random motion of the human.

3.8.2.2 HFR trajectory evaluation

The human tracking is performed on the different algorithms and baselines, and trajectory of the human and the robot is plotted on all approaches. The results are given in Figure 3.10 for Scenario ID I, Figure 3.11 for Scenario ID II, and Figure 3.12 for Scenario ID III. In all the scenarios, the proposed social tracker gave the best results as compared to all other approaches and baselines.

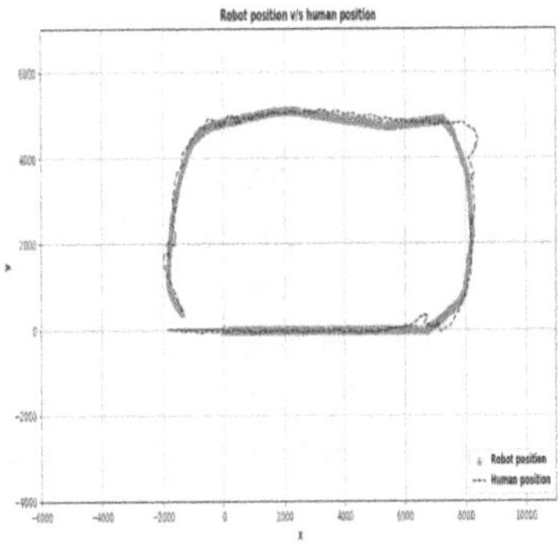

(a)

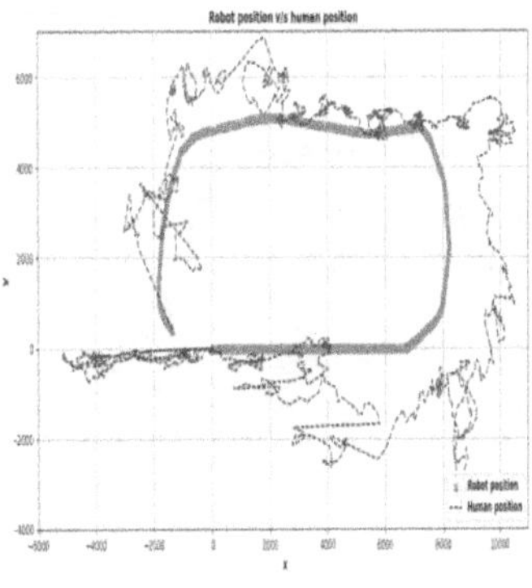

(b)

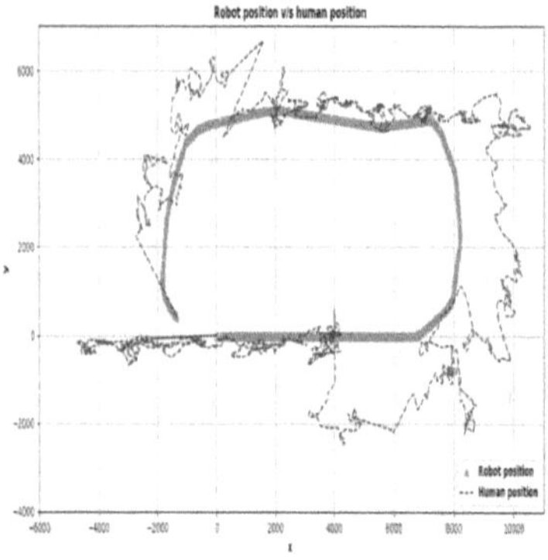
Robot position v/s human position
Robot position
Human position

(c)

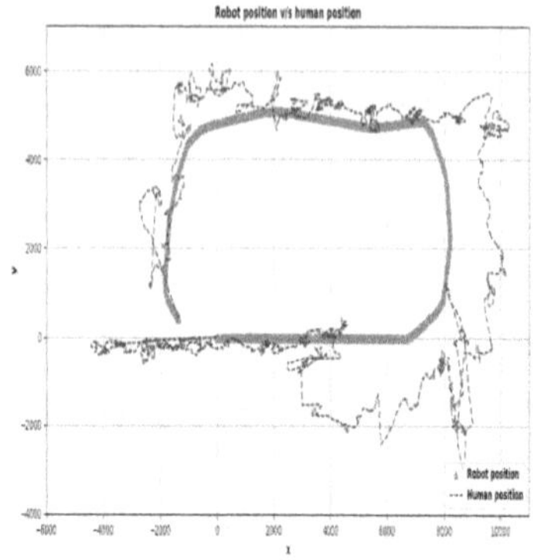
Robot position v/s human position
Robot position
Human position

(d)

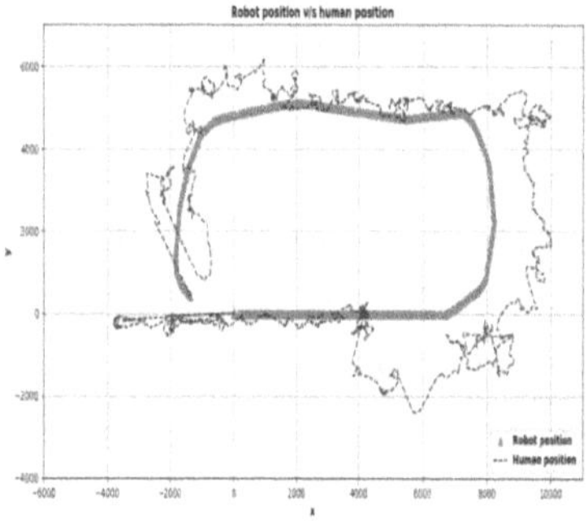
Robot position v/s human position
Robot position
Human position

(e)

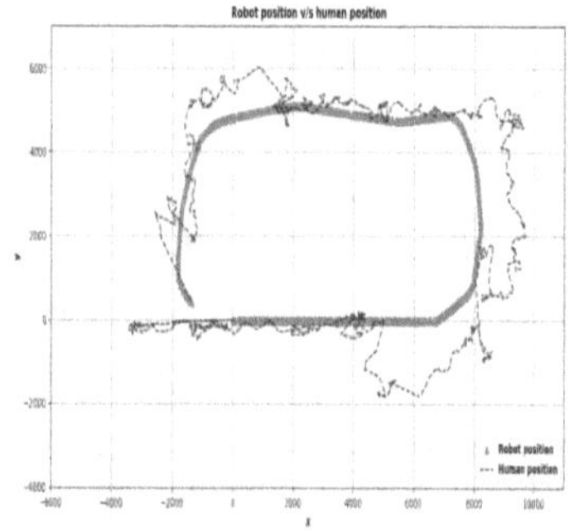
Robot position v/s human position
Robot position
Human position

(f)

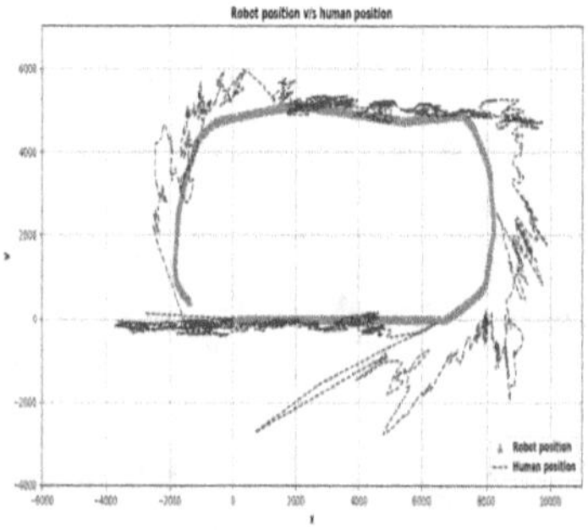
Robot position v/s human position
Robot position
Human position

(g)

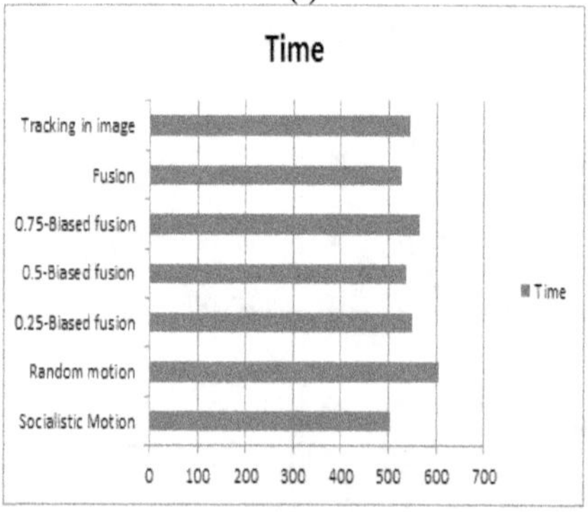
Time
Tracking in image
Fusion
0.75-Biased fusion
0.5-Biased fusion
0.25-Biased fusion
Random motion
Socialistic Motion
0
100
200
300
400
500
600
700
Time

(h)

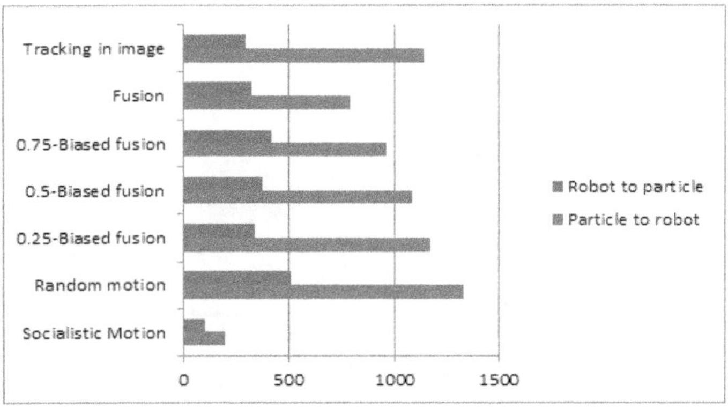

(i)

Figure 3.10: Scenario ID I comparison (a) Social Motion (b) Random motion (c) 0.25-biased motion (d) 0.5-biased motion (e) 0.75-biased motion (f) Fusion (g) Tracking in image (h) Computation Time (i) Error

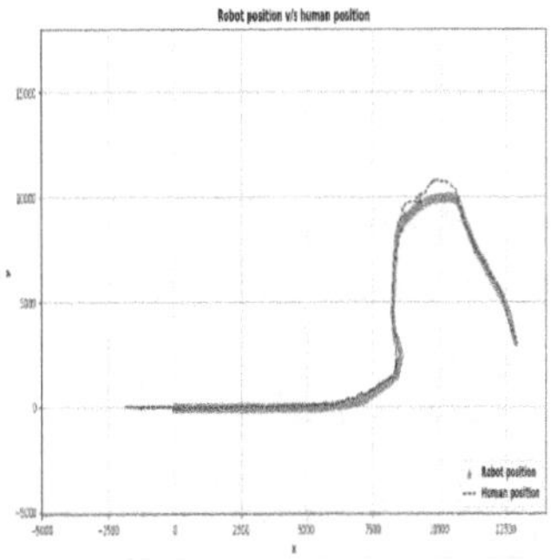

(a)

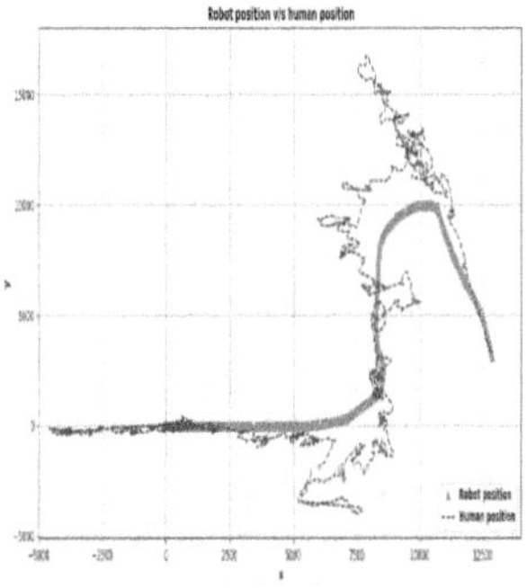

(b)

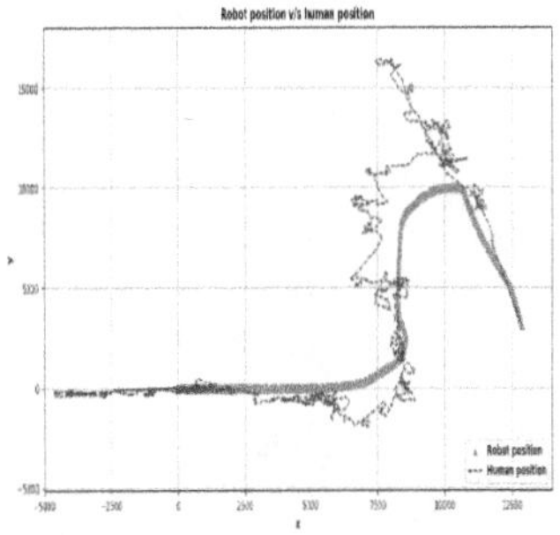

(c)

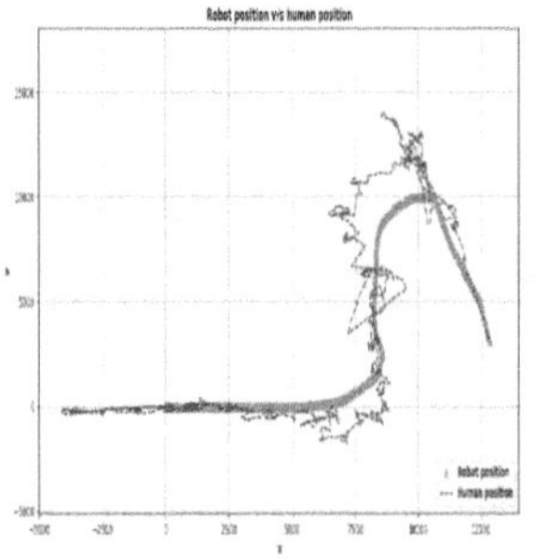

(d)

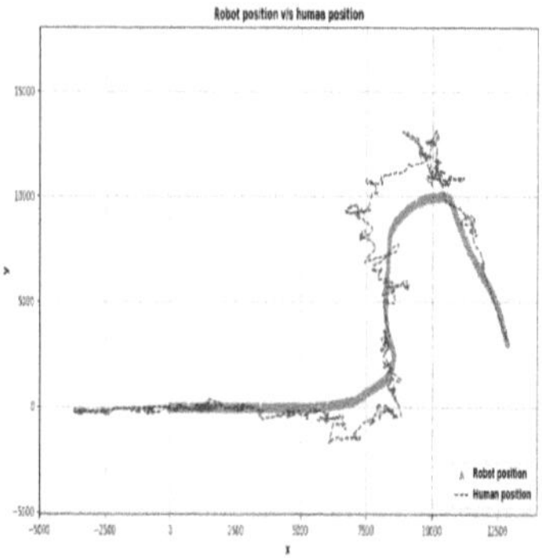
Robot position v/s human position
Robot position
Human position

(e)

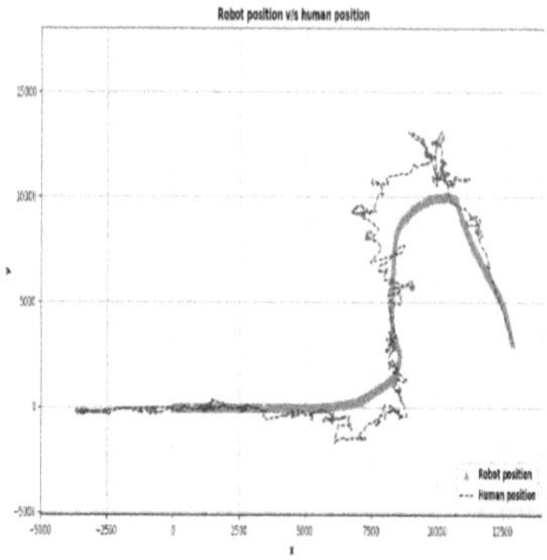
Robot position v/s human position
Robot position
Human position

(f)

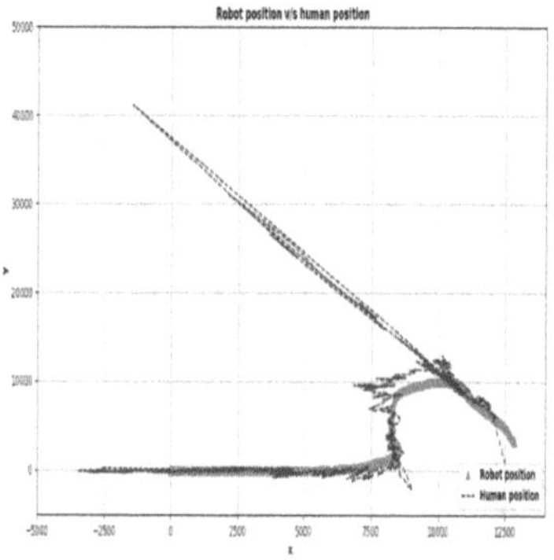
Robot position v/s human position
Robot position
Human position

(g)

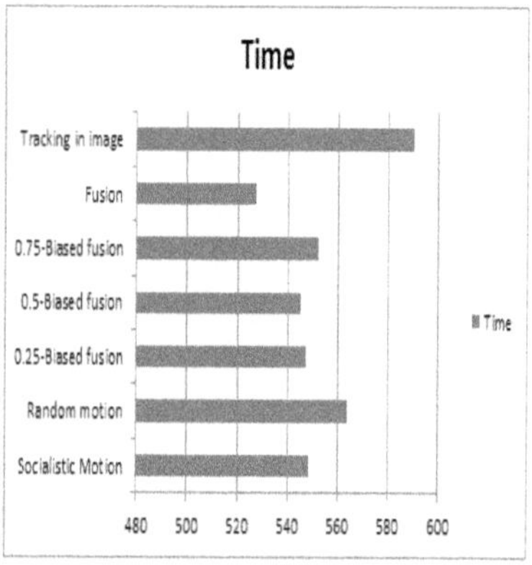
Time
Tracking in image
Fusion
0.75-Biased fusion
0.5-Biased fusion
0.25-Biased fusion
Random motion
Socialistic Motion
480
500
520
540
560
580
600
Time

(h)

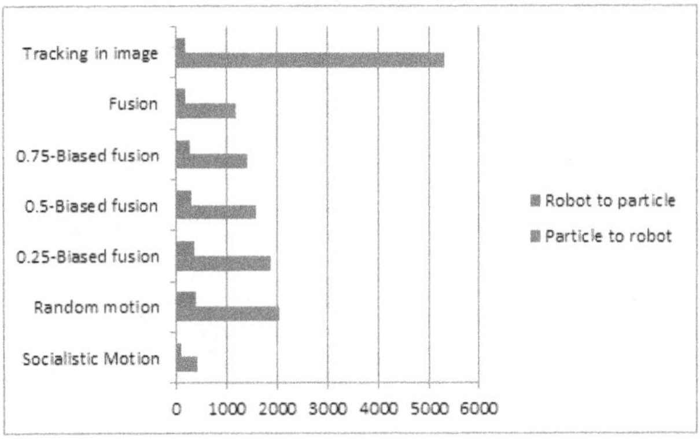

(i)

Figure 3.11: Scenario ID II comparison (a) Social Motion (b) Random motion (c) 0.25-biased motion (d) 0.5-biased motion (e) 0.75-biased motion (f) Fusion (g) Tracking in image (h) Computation Time (i) Error

Table 3.2: Comparison with different number of particles

Number of particles	d_{trace} (mm)	$\overline{d_{trace}}$ (mm)
50	190.82	97.30
100	193.79	101.56
150	196.03	101.99
250	199.41	102.72
300	198.64	103.34
350	196.72	102.05
400	198.91	103.10
450	199.08	102.29
1000	199.17	103.60
2000	200.65	105.08

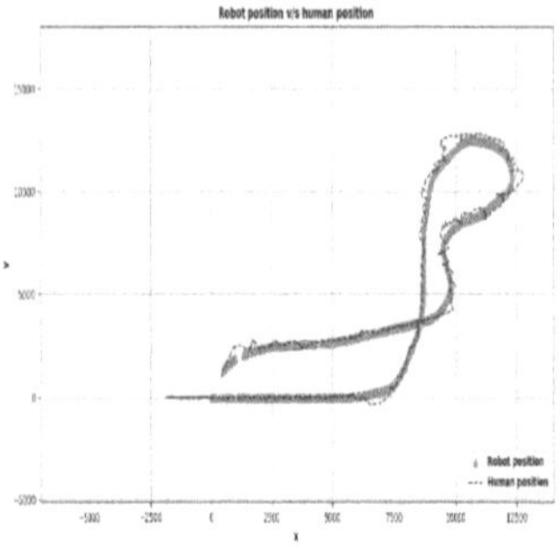
Robot position v/s human position
Robot position
Human position

(a)

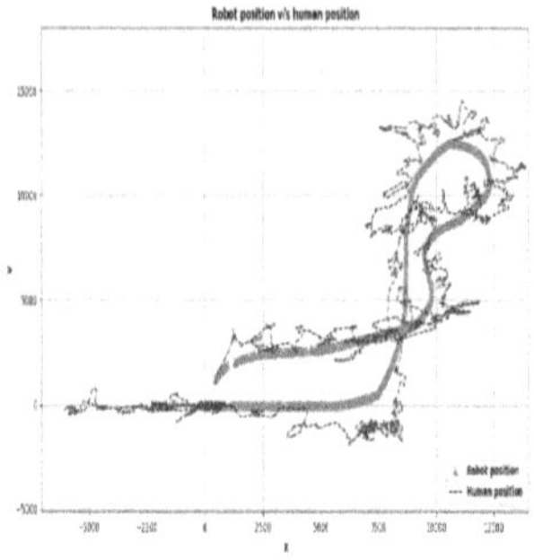
Robot position v/s human position
Robot position
Human position

(b)

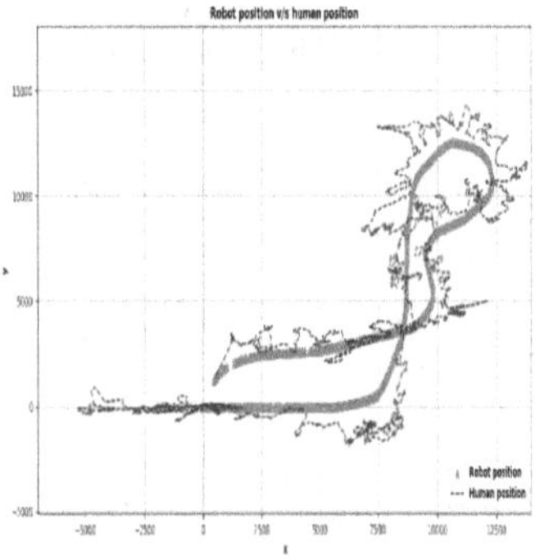
Robot position v/s human position
Robot position
Human position

(c)

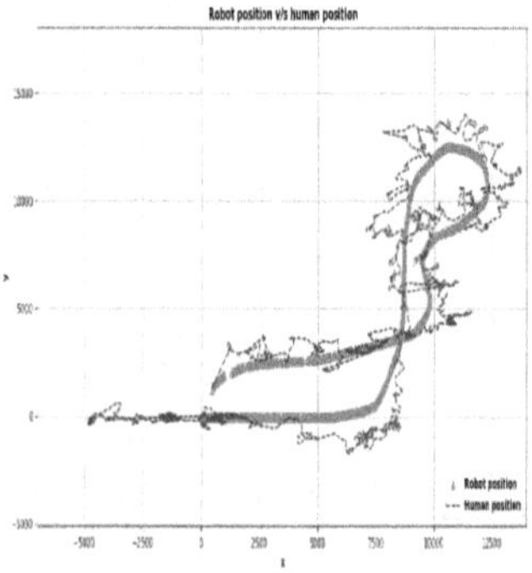
Robot position v/s human position
Robot position
Human position

(d)

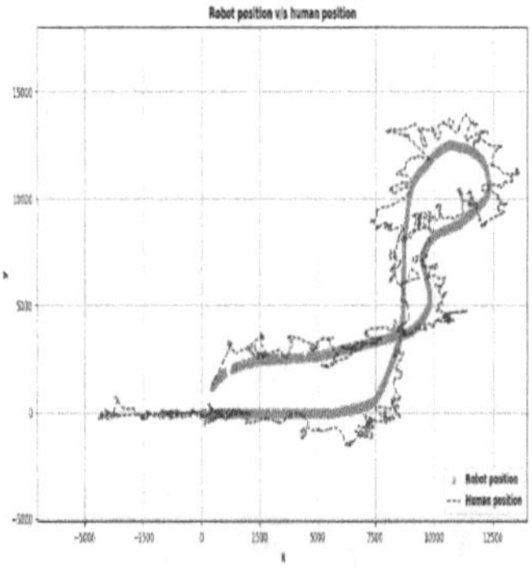
Robot position v/s human position
Robot position
Human position

(e)

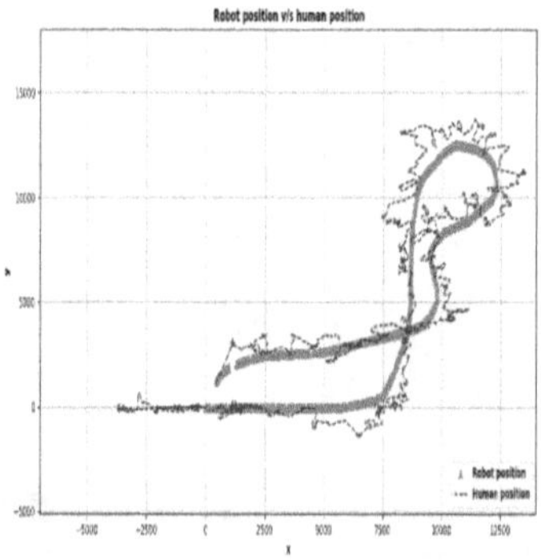
Robot position v/s human position
Robot position
Human position

(f)

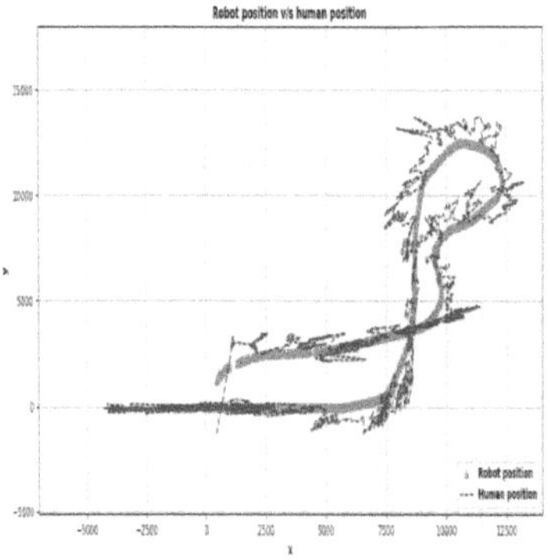

(g)

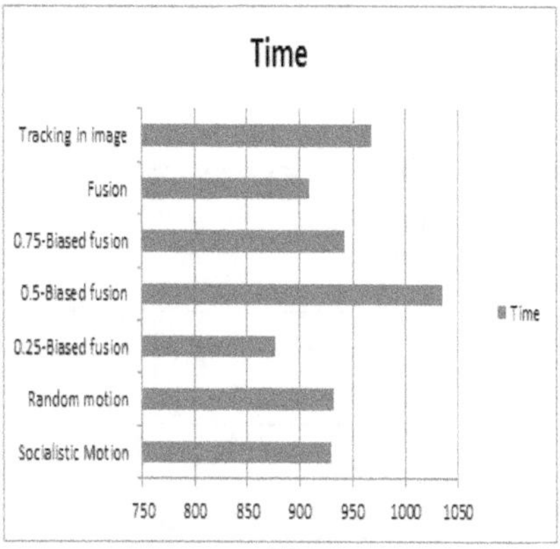

(h)

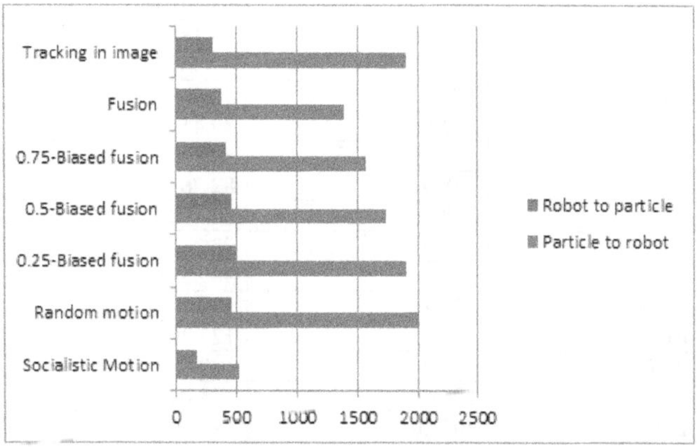

(i)

Figure 3.12: Scenario ID III comparison (a) Social Motion (b) Random motion (c) 0.25-biased motion (d) 0.5-biased motion (e) 0.75-biased motion (f) Fusion (g) Tracking in image (h) Computation Time (i) Error

3.8.2.3 HFR parameter analysis and evaluation

This thesis also gives parameter analysis on the distinct parameters such as motion noise, variance and number of particles. It is observed that the smaller number of particles give the best performance, which indicates the strength of the model. As a result, a large number of particles to increase diversity are not necessary. Table 3.2 summarises the error with respect to the number of particles. Table 3.3 studies the observation variance parameter (σ^2). The ideal values are the ones that capture the diversity of observations within the image as the uncertainty increases. Finally, the testing with the motion noise is set out in Table 3.4. The noises also correspond to the person generally following the robot religiously and hence a small noise is enough to model the following behaviour uncertainties.

Table 3.3: Comparison with different observation variance

Observation Variance σ^2	d_{trace} (mm)	$\overline{d_{trace}}$ (mm)
0.125	270.78	220.10
0.5	265.73	213.38
12.5	269.30	217.06
50	220.58	126.21
200	209.37	106.83
800	200.72	102.41
1800	193.70	99.36
3200	187.21	98.13
4050	185.60	96.18
6050	179.18	92.55
8580.5	175.17	90.64
11400.5	172.69	89.45
20200.5	164.39	84.81
31500.5	158.91	82.76
61250	158.90	86.49
86112.5	162.51	93.23
125000	166.56	98.83
245000	175.22	112.42
405000	185.11	123.20
500000	190.04	129.83
605000	196.40	138.04
819200	205.86	145.49

3.8.2.4 Discernibility of 'HFR' in real-time

A few snapshots of the 3D tracking for scenario ID I are shown in Figure 3.13. Here the robot took a single tour of the laboratory and the human subjects followed the robot. In this scenario the robot took a rotation four times at four different corners.

This thesis also performed human tracking in another scenario with ID II. Here the robot took a few rotations and tracked the human at several paces and corners. The tracking is shown at Figure 3.14.

Table 3.4: Comparison with different motion noise variance

Motion noise	d_{trace} (mm)	$\overline{d_{trace}}$ (mm)
0.0	260.41	194.18
0.02	247.02	186.28
0.04	222.42	162.68
0.07	190.61	125.73
0.08	179.44	112.61
0.10	169.85	94.64
0.15	172.04	89.43
0.20	184.46	95.23
0.25	191.41	98.37
0.34	200.59	103.82
0.39	202.30	104.40
0.44	207.23	106.62
0.5	208.92	106.12
0.78	216.22	107.35
1.04	215.34	106.20
1.57	229.62	110.05
1.81	274.02	137.38
2.11	411.04	219.39
2.5	611.88	333.14

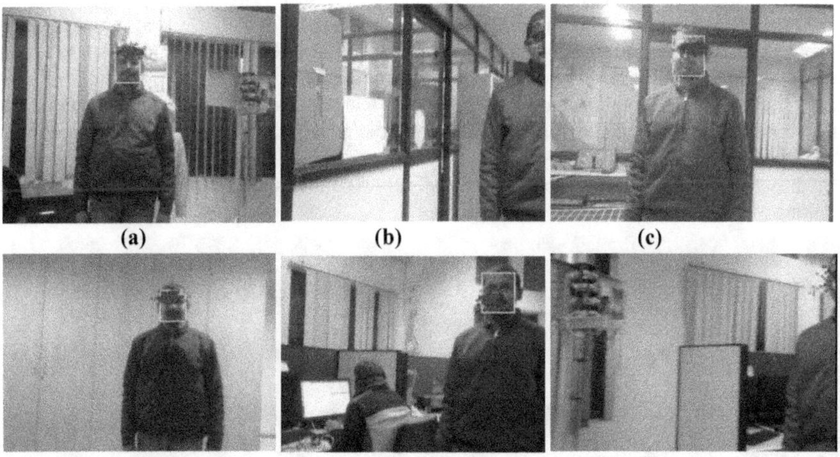

(a) (b) (c)

(d) (e) (f)

Figure 3.13: Tracking behavioural module on scenario ID I (a) robot is at an initial position and a human is detected and tracking started. (b) Robot reached the corner and took a left turn, although face of the human was not detected, tracking happened successfully. (c) Robot took anticlockwise (90°) rotation and moved forward and tracking of the human continued. (d) Robot again took anticlockwise (90°) rotation and tracked the human who was following the robot. (e) Robot again reached a turning point and took a left turn, and consequently tracked the human behind it. (f) Robot at situation where the whole face of the human was not captured by the monocular camera, so detection of face is not possible, but the tracking module successfully tracked the human face.

(a) (b) (c)

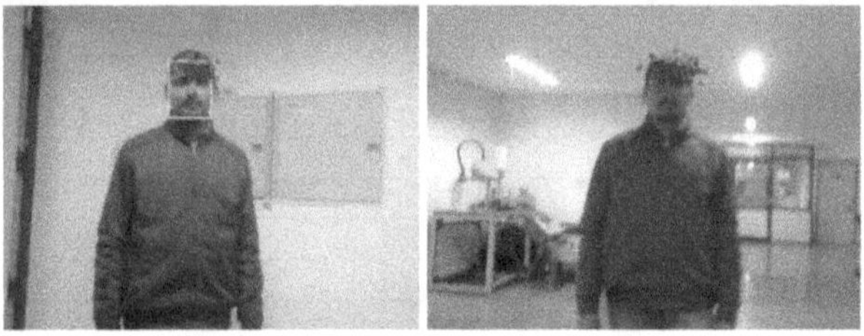

(d) (e)

Figure 3.14: Tracking behavioural module on scenario ID II (a) Robot was at initial position and human was detected and tracking started. (b) Robot reached at the corner and took a left turn, although face of the human was not detected, tracking of human continued successfully. (c) Robot took anticlockwise (90°) rotation and moved forward. Detection and tracking of human continued. (d) Robot took a right turn and human was following the robot. (e) Robot again took a clockwise (90°) rotation and human was following the robot continuously, here face of the human was not detected by the robot but tracking module robustly tracked the human at the corner when robot was taking a turn.

(a) (b) (c)

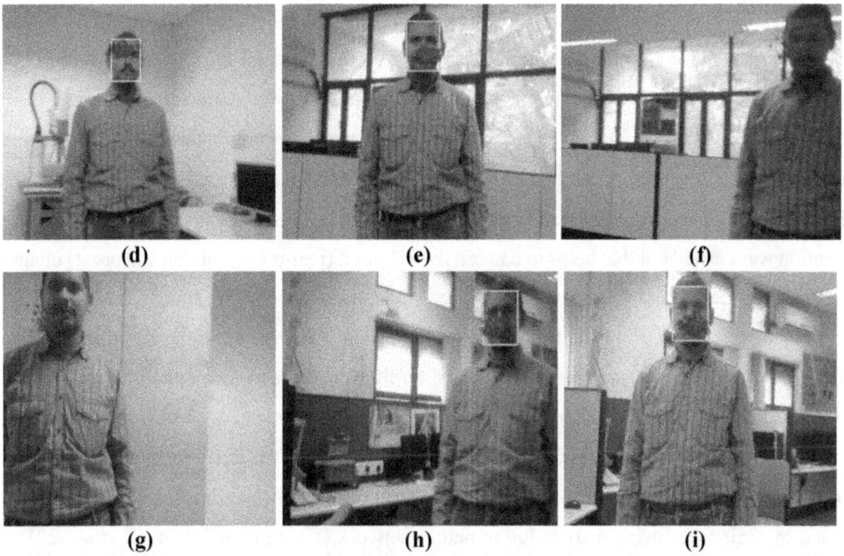

Figure 3.15: Tracking behavioural module on scenario ID: III (a) Robot covered a short distance and human was following the robot, since robot was moving horizontally in the same direction so detection and tracking both were performing well. (b) Robot took anticlockwise (90°) rotation at the corner and detection and tracking of human continued. (c) Robot gradually took a right turn so both (detection and tracking) algorithm performed well. (d) Robot took clockwise (90°) and moved forward and human was following the robot maintaining a comfortable distance, so detection and tracking both continued. (e) Robot again took clockwise turn and human was behind the robot at a comfortable distance, so robot was able to detect and track human. (f) Robot suddenly took a right turn with certain angle, here face detection method was not able to detect human's face but robot was easily handling this failure condition of the detection method and skilfully tracked the human at the corner. (g) Robot just took clockwise (90°) rotation at the corner and started moving. Here detection was not successful, but the tracking approach tracked the human. (h) Robot was gradually taking anti-clockwise rotation and moved simultaneously, so detection and tracking both methods performed well. (i) Robot completed its journey and human was behind the robot while maintaining comfortable distance from the robot, so detection and tracking both modules continued working.

Scenario ID III is more complicated because the robot took several turns as the human followed. The robot tracked the human at each rotation and corner while the human got out of view on multiple occasions. The pictorial demonstration of scenario ID III is shown in Figure 3.15.

3.8.3 Multi-behaviour 3D-PSF model

This proposed work is also tested on real-life settings at the Centre of Intelligent Robotics, IIIT Allahabad with the AmigoBot robot and a rear-facing webcam, where a few lab members were asked to casually follow the robot, while behaving erratically at times. The Amigobot robot was placed at a pre-determined position. The camera was mounted on top of the robot. The AmigoBot robot was used to track humans. The AmigoBot robot was stationed at the origin (0,0) and the humans stand behind the robot at a certain distance. The coordinates of the humans are taken as the (x, y) position of the humans and the height of humans is used as the Z coordinate. The transformation between the robot base and the camera was manually measured. Camera calibration was carried out that helps to convert the 3D point (real-plane) into the 2D point (image-plane). Thereafter, different processes run nearly independently on the same system. The first process is the teleoperation of the robot. The teleoperation is carried out by using robot-specific libraries. As the robot moves, a service provided by the vendor keeps computing the position of the robot. The position is computed by using the high-precision encoders on the robot. Finally, a separate module runs the particle filter proposed in this chapter. The development is carried out using the Python programming language. The camera continuously takes images. The humans are detected by their faces using a convolution neural network (Zhang et al., 2016; Li et al., 2015). The initial position of the particles is computed using an inverse transform of image coordinates to a person's position for each person. Thereafter, the program runs the Particle Filter logic inside a loop. At every iteration, the camera takes a new image as input and the face detection is carried out. The particles are moved based on the position of the robot computed from the encoders and the published positions of the different persons. If the person is detected, the observation model is applied. The deletion of track and the re-initialisation happens by handling the different timers set for the purpose as per the logic. Every human in a group usually follows the robot and due to their social behaviour when any human leaves its group, then the track is deleted for that human. In the continuation of the navigation process as the human joins their group, the human is detected in the 2D image plane. This 2D image plane coordinate is converted into 3D real-plane using inverse camera calibration and inverse transformation. This 3D coordinate is the re-initialised point of the human and re-assigned to the tracker for further tracking. After all computations, the weighted addition of all particles is used to compute the position of the person, and the same is logged and published.

The real-life experiments do not have the ground truth of the tracked persons. Therefore, the experiments were also carried out on simulated robots where one robot enacted the real physical robot with a simulated rear-facing web-camera, and the other robots enacted by the following people. These robots were teleoperated and the teleoperating humans were asked to maintain as many social norms as possible. To make the simulations realistic, the map used was that of the same place in which the social experiments were made. The 2D map was made using a SLAM library using the Pioneer LX robot. The 3rd dimension was added by assuming the obstacles have a fixed common height. The simulated camera was restricted to the same constraints as the real camera, using the same projection matrix. The simulated camera was programmed so that the visitors occluded by static obstacles are not detected. The simulated camera was noised to imitate the large errors, including errors in detection and recognition, associated with cameras. This thesis used the Robot Operating System (ROS) library as a generic framework and MobileSim simulator for the Amigobots. First, the environment map was loaded on the MobileSim simulator. Four robots were placed in the MobileSim simulator. RosAria library was used which provides the basic functionalities of the simulated robots including integration with ROS. All four robots have their coordinates, and one robot was made to behave like a leader robot and the remaining three robots acted like humans. Robots were operated by teleoperation. Here four different computing systems were needed to teleoperate the four robots (1 imitating the leader robot and 3 imitating the humans). One system was made as a ROS master system (that imitated the leader robot) and the remaining three systems connected to the ROS master system. The proposed model knows the actual trajectory of all robots from the simulator and this trajectory was used as the ground truth for comparing the proposed motion model. The trajectory was continuously logged. The same program of the particle filter was used, except that the camera image was taken by the simulated camera. The performance of the 3D PSF algorithm is benchmarked using the key performance indicators considering the specific database challenges for multiple persons in a group. First, the camera is a monocular and low cost with a limited field of view so 1068 frames out of 5404 are affected by noise due to sun light and lightning condition, leading to a lack of detection While robot is navigating in the workspace it also takes several rotations to accomplish its goal and therefore the humans' visibility gets lost 28 times for 129 seconds due to the restricted field of view of the camera. The aim is to measure the algorithm accuracy in a dataset containing erratic human behaviours. Therefore, the dataset is recorded with the erratic behaviours of overtaking (4 occurrences), swapping positions (9 occurrences), visitors following another human (5 occurrences), leaving the group (9 occurrences) and re-joining the group (9 occurrences).

Table 3.5: Parameters values

SN	Parameters	Value	Remarks
1.	Number of particles	200	Set by trial and error
2.	Robot radius	0.2533 m	As per actual measurement
3.	Person radius	0.2533 m	As per actual measurement
4.	Average velocity of the human	1.500 m/s	As per actual measurement
5.	Height of the camera	1.220 m	As per actual measurement
6.	Calibration matrix	$\begin{pmatrix} 718.46 & 0.0 & 314.62 & 0.0 \\ 0.0 & 719.49 & 265.41 & 0.0 \\ 0.0 & 0.0 & 1.0 & 0.0 \end{pmatrix}$	Obtained from calibration
7.	Image resolution	(640×480)	As per actual
8.	Coefficient of person to robot attraction force (w_1)	1.3181	Set so that the attraction force is at a set threshold when the person is at a pre-determined social distance from the robot
9.	Coefficient of person to robot repulsion force (w_2)	2473.082	Set so that attraction and repulsion forces cancel each other when the person is at a pre-determined social distance from the robot
10.	Coefficient of person to person attraction force (w_3)	2.0251	Set so that the attraction force to the person is 1/3rd of the attraction force to the robot when the person is at a pre-determined social distance from the robot and another person
11.	Coefficient of person to person repulsion force (w_4)	824.361	Set so that attraction and repulsion forces cancel each other when two persons are at a pre-determined social distance from each other
12.	Coefficient of random force (w_5)	1500.0	Set to make the force in the same range as the other forces
13.	Coefficients related to the total force	W_{th}=0.9, W_{min}=0, W_{max}=0.2;	As per program logic
14.	Minimum angle to qualify a turn (ð), as a Cosine value	0.6	Set by measuring the angle that the robot makes at different turns in the environment
15.	Minimum time of invisibility to delete a track (Δ)	4 secs.	Set by measuring the actual time that the humans left in the experiments due to invisibility due to genuine reasons apart from corners

This thesis initially discusses the comparisons with the simulated setup and then discusses the experiments with the real-life humans under social settings. All the behaviours are implemented in several scenarios and further the proposed model is employed onto those different scenarios. Here three scenarios are presented. The general behaviour is the humans following the robot for their navigation. The robot knows only the initial position of the humans and anticipates the humans' trajectory in both the 3D real world and the 2D image plane.

One of the important aspects is the parameter setting. Some parameters were directly measurable like the radial occupancy of the robot and the person, the average walking speed of the humans, and the camera calibration matrix. There are numerous coefficients related to the forces and trial and error between all the coefficients to set an appropriate value is not possible. Hence, first the distance that a human prefers to keep from the robots was measured by doing some social experiments. At this social distance value, it is known that the attraction and repulsion forces between the person and the robot should cancel each other out. The proposed model arbitrarily fixes the attraction force coefficient value at this distance and sets the repulsion force coefficient value to cancel the attraction force. Similarly, the proposed model also measured the social distance that two people tend to keep from each other. Similar equilibrium conditions were considered between two people and the robot. Heuristically, the proposed model set the attraction between two people to be $1/3^{rd}$ in contrast to the attraction between a robot and a person. This is based on the observation that while following, the people are primarily attracted towards the leader with a small attraction towards each other. The repulsion force coefficient is set such that the repulsion force cancels the attraction force between people at the social distance. The random force is scaled to the same scale. This thesis performed several experiments to decide the number of particles. In general, a higher number of particles are preferable, which however also increases the computation time. The parameter values are described in Table 3.5.

3.8.3.1 Comparisons of Multi-behaviour 3D-PSFM

To benchmark the dataset, this thesis used 4 baseline approaches. The first baseline approach is a particle filter with a random motion model. The filter deletes a track when the person is non-observable for a long time and re-initialises the track when the person is visible again. The second approach directly tracks the subjects in the image plane. However, knowing the height of the person, the conversion of the position from the 2D image to a 3D position in real-world is additionally carried out for comparisons. The next method is the Kalman Filter.

The other competent methods are from the trajectory prediction approaches. The LSTM based implementations are increasingly becoming popular. In the experiments, the assumption is that the humans are only partially observable when they are within the field of view of the camera, and hence the training data is highly limited not covering most of the interesting cases when the human is outside the field of view. Even if it is critiqued that external sensors can be used to make training data, the proposed model assumes that the humans have unclear intents and a stochastic behaviour. Hence in the same situation, the person could react differently. Moreover, the persons occasionally leave and re-join groups which is a displayed behaviour but should not be learnt. Initially, an attempt was made to see if the LSTM can reject such noises and learn using simulated data. However, training on one person's data performed poorly when tested on the data of another person. This also highlights a known observation that different people behave differently. Hence, training was repeated when training and testing on the same person's data. The LSTM took the positions of the robots and all persons to predict the motion of the person under study.

An unfair disadvantage of the LSTM was that the other methods benefited from the continuous observations from the camera, while the LSTM only used the initial conditions along with the known position of the robot. Hence another method was used where the LSTM was actively re-initialised from the position calculated from the image. The LSTM was used to predict a short sequence. Thereafter, the LSTM was fed a small trajectory computed from the detections in the image, and the LSTM was made to predict another short sequence.

Table 3.6: Comparisons of the proposed approach

Scenario ID	Person ID	Proposed (mm)	Random Particle filter (mm)	Tracking in Image (mm)	Kalman Filter (mm)	LSTM (with re-initialisation) (mm)	LSTM (without re-initialisation) (mm)
I	1	**629.28**	662.26	4135.63	998.69	1950.71	2249.56
	2	**618.69**	670.93	4410.85	985.54	3147.95	3448.09
	3	**258.24**	601.85	5128.56	958.67	3400.55	3682.45
II	1	**234.88**	303.87	10577.32	802.64	4802.54	5604.49
	2	**229.96**	284.24	1767.64	706.27	8638.65	9129.27
	3	**388.56**	468.88	2894.84	1508.49	8612.41	8932.18
III	1	**406.27**	738.69	7212.76	959.64	3126.22	6702.37
	2	**676.32**	739.63	30610.39	1202.34	5465.19	8312.73
	3	**385.18**	810.94	6215.89	1284.75	9265.74	7798.63

Table 3.6 gives the error calculated for the humans' actual and tracked trajectory in 3D. The metric of comparison is the root mean squared error (RMSE) between the actual and the

tracked positions of the humans. Here RMSE is the key performance indicator. Each algorithm has an option of denial of service where it deletes the track assuming that the person has left. While calculating the RMSE, a penalty of 1400 is added for all cases when the algorithm wrongly predicted a person leaving. Otherwise, the algorithms can decline service for the whole motion to get the least errors. This thesis also gives a statistical analysis of all the experiments reported. A Pairwise T-test was used to understand the statistical significance of the proposed approach with all the other approaches. Based on statistical testing it was observed that the proposed algorithm is better than all other methods used for comparisons with a significance level of at least 95%.

Table 3.7: F-score for detection of human leaving the group

Scenario ID	Person ID	Proposed (with corner detection)	Proposed (without corner detection)
I	1	**0.98**	0.07
	2	**0.98**	0.46
	3	**0.94**	0.09
II	1	**0.96**	0.81
	2	**0.98**	0.78
	3	**0.96**	0.79
III	1	**0.97**	0.68
	2	**0.98**	0.55
	3	**0.98**	0.61

Table 3.8: Confusion matrix

Scenario ID	Person ID	Proposed (with corner detection)				Proposed (without corner detection)			
		TP	TN	FP	FN	TP	TN	FP	FN
I	1	**98**	**10827**	**0**	**42**	90	8777	2020	80
	2	1026	**9911**	**0**	**30**	**1079**	7392	2425	71
	3	**960**	**9897**	77	**33**	125	8589	2174	79
II	1	**3138**	**12541**	**0**	**200**	3128	11364	1177	210
	2	**2247**	**13552**	**0**	**80**	2245	12389	1163	82
	3	2967	**12712**	**0**	200	**3018**	11303	1409	**149**
III	1	**2500**	**22659**	**2**	**120**	2495	20466	2193	125
	2	2283	**22916**	**2**	80	**2285**	19388	3528	**78**
	3	**2142**	**23059**	**0**	**80**	2135	20477	2580	87

This thesis tested the proposed method on three different scenarios to ensure the validity of the proposed 3D particle model. The proposed model gives errors between 0.2 metres to 0.6 meters which are smaller as compared to the other methods. The Random particle filter method cannot be applied everywhere because it gives a large number of errors for Scenarios I and III, while the errors are lower in Scenario II, still higher than the proposed approach. The performance of the random particle filter deteriorates when the person is not visible, while the filter soon erroneously deletes the track while the robot may have been turning a corner. When tracking in the image, the errors are extremely high because of the moving camera. The errors increase as soon as the person is outside the field of vision of the robot, in which case the position is guessed under the assumption of smoothness in the image, which is void considering that the robot was actually making a sharp turn and a stationary human shall also result in a sharp motion in the image, which the filter erroneously smooths out. The Kalman Filter performs poorly due to a linearity assumption that does not hold good in this model. As the robot rotates, the non-linearity significantly increases thus increasing the errors. It must be stressed that the motion model in the proposed approach is based on heuristics that act stochastically and hence there is no easy way to get the derivative for the implementation of the Extended Kalman Filter.

Table 3.9: Probabilistic interpretation

Scenario ID	Person ID	Proposed (with corner detection)			Proposed (without corner detection)		
		Precision	Recall	Accuracy	Precision	Recall	Accuracy
I	1	**1.00**	0.70	**0.99**	0.04	0.30	0.05
	2	**1.00**	**0.97**	**0.99**	0.52	0.93	0.61
	3	**0.92**	**0.96**	**0.98**	0.80	0.77	0.79
II	1	**1.00**	**0.94**	**0.98**	0.72	0.93	0.91
	2	**1.00**	**0.96**	**0.99**	0.65	**0.96**	0.92
	3	**1.00**	0.93	**0.98**	0.68	**0.95**	0.90
III	1	**0.99**	**0.95**	**0.99**	0.53	**0.95**	0.90
	2	**0.99**	**0.96**	**0.99**	0.39	**0.96**	0.85
	3	**1.00**	**0.96**	**0.99**	0.45	0.93	0.89

The LSTM performed poorly because it also learnt the noisy behaviours like leaving the group, getting distracted, etc. In the proposed assumption, humans often act erratically, and those behaviours cannot be filtered out. Training LSTM on the entire subject-specific makes it impossible for LSTM to decide which are the erratic behaviours and which are the genuine behaviours, and the performance is thus reasonably poor. After re-initialisation, the LSTM performance improved reasonably well. However, overall, the performance was still not good due to the same reasons. The initial position given to the LSTM was itself noisy since it came from a noisy camera. Observing such noisy sequences gave a reasonably large initial error to the LSTM. The LSTM was not trained on such data and thus continued to perform even more poorly. It must be stressed that the thesis does not intend to show that the LSTM is ineffective for solving the problem. The intention behind the comparison of the proposed method is to show that the LSTM fails to perform under the settings of noisy sensor readings and highly noisy data where people often do not follow the robot. Under ideal settings, the LSTM could have given a reasonable performance.

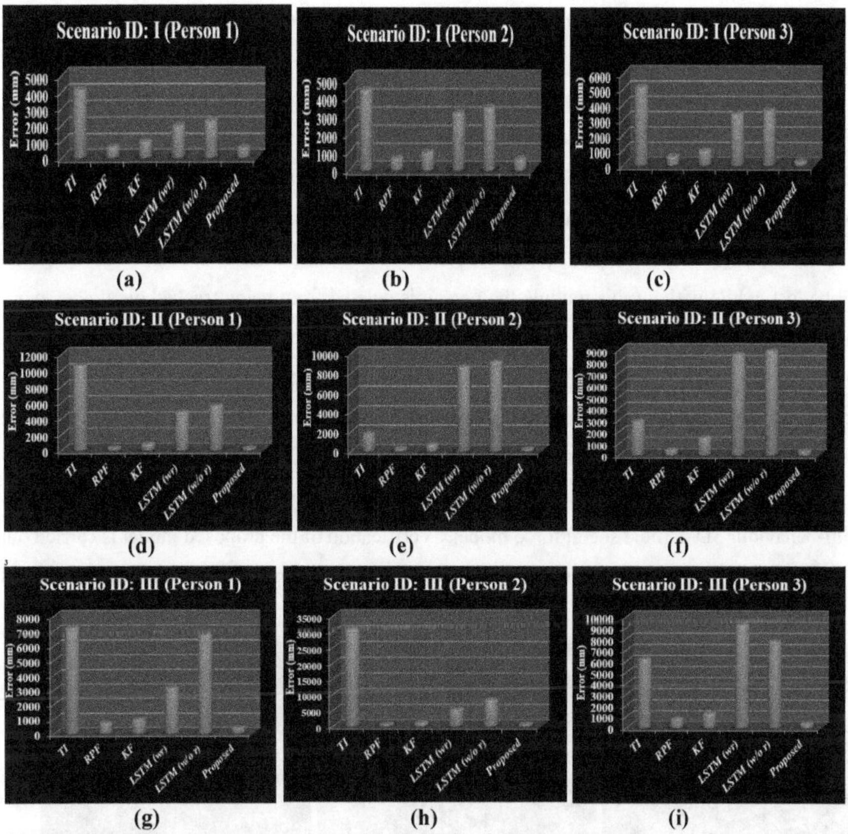

Figure 3.16: Root mean square error (TI: Tracking in Image, RPF: Random Particle Filter, KF: Kalman Filter; LSTM (wr): LSTM (with re-initialisation), LSTM (w/o r): LSTM (without re-initialisation)) (a) Scenario ID: I Person 1 (b) Scenario ID: I Person 2 (c) Scenario ID: I Person 3 (d) Scenario ID: I Person 1 (e) Scenario ID: II Person 2 (f) Scenario ID: II Person 3 (g) Scenario ID: III Person 1 (h) Scenario ID: III Person 2 (i) Scenario ID: III Person 3

The thesis proposed a mechanism to detect the robot turning and use the same to anticipate the person moving out of the field of sight, in which case the track was not deleted despite a loss of visibility for a prolonged time. Overall, the detection of a person leaving the group was an integral part of the algorithm and hence its performance is also benchmarked. Comparisons are done with the naïve implementation of the particle filter that does not anticipate the loss of visibility due to corners and turns with the additional challenge of erratic Human Behaviours. An additional key performance indicator is the ability to detect people leaving the group measured using the standard binary classifier metrics. F-score is used for detecting people leaving the group and the results are presented in Table 3.7.

The results suggest that the proposed method gives significantly better results using the corner detection module. Here the count of true positives (TP) and true negatives (TN) is increased and due to these false negatives (FN) have been reduced as shown in Table 3.8. Specifically, false positives (FP) have rapidly decreased and close to zero as compared to without corner detection. The proposed algorithm anticipates that since the robot is making a turn at the corner, the visibility will be lost, while without this module the person is deleted from being tracked on almost every corner. It may be counterintuitive that this does not result in an increase in false negatives, which is because the misclassification is a rarity. The little increase in true positives and true negatives; and the reduction in false positives is attributed to the accuracy of tracking. Furthermore, in Table 3.9 the precision, recall, and accuracy have also improved. Accuracy is 98% and above in each scenario. The precision is also high and these interpretations also prove the validity of the proposed multi-behaviour 3D particle social force model. Verification of the proposed model is carried out by the actual trajectory of the multi-agent system where the proposed model knows the ground truth and the humans leaving is manually annotated.

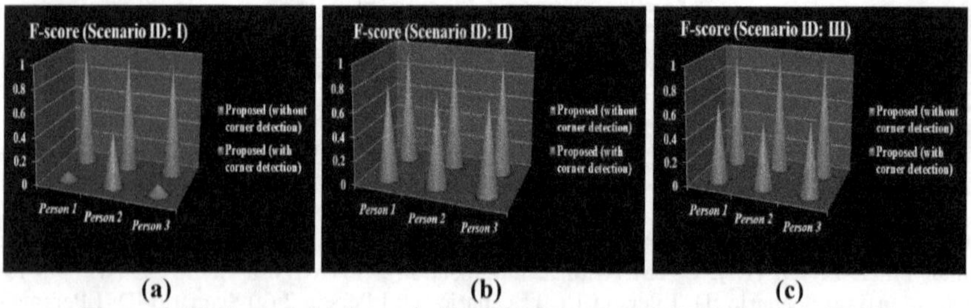

(a) (b) (c)

Figure 3.17: F-score (a) Scenario ID: I (b) Scenario ID: II (c) Scenario ID: III

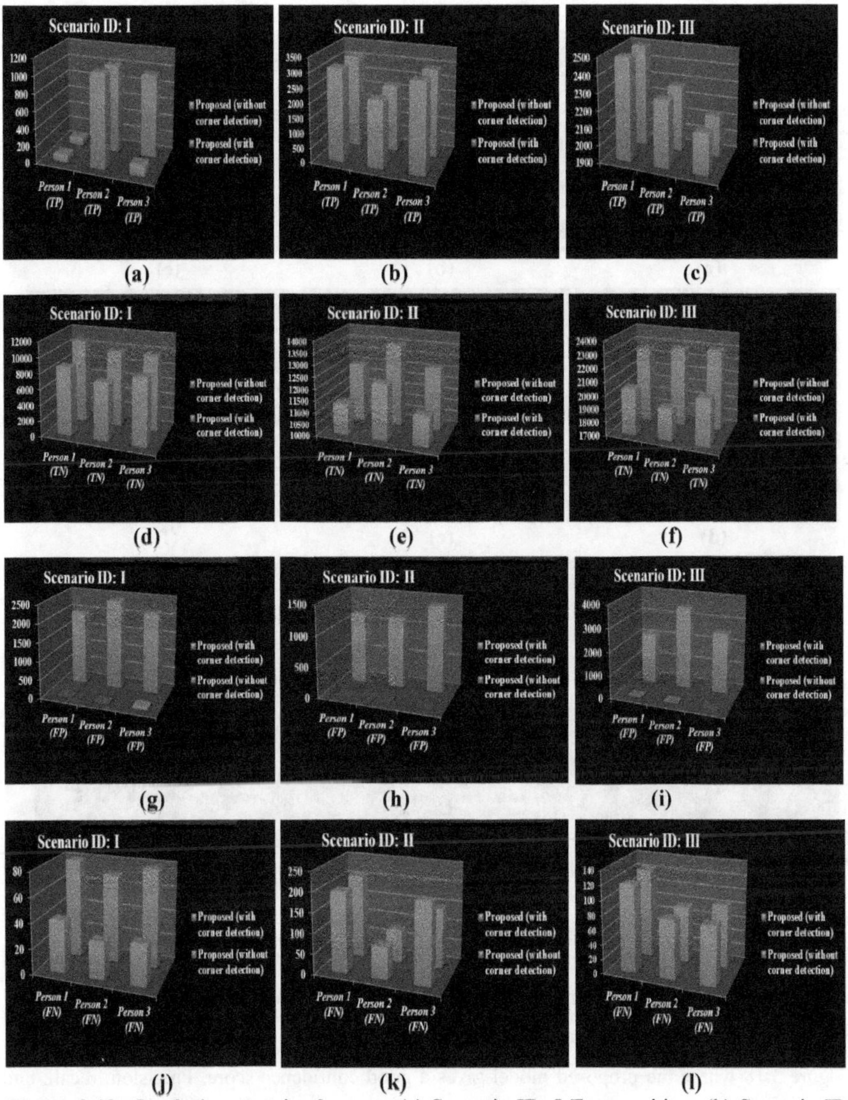

Figure 3.18: Confusion matrix elements (a) Scenario ID: I True positives (b) Scenario ID: II True positives (c) Scenario ID: III True positives (d) Scenario ID: I True negatives (e) Scenario ID: II True negatives (f) Scenario ID: III True negatives (g) Scenario ID: I False positives (h) Scenario ID: II False positives (i) Scenario ID: III False positives (j) Scenario ID: I False negatives (k) Scenario ID: II False negatives (l) Scenario ID: III False negatives

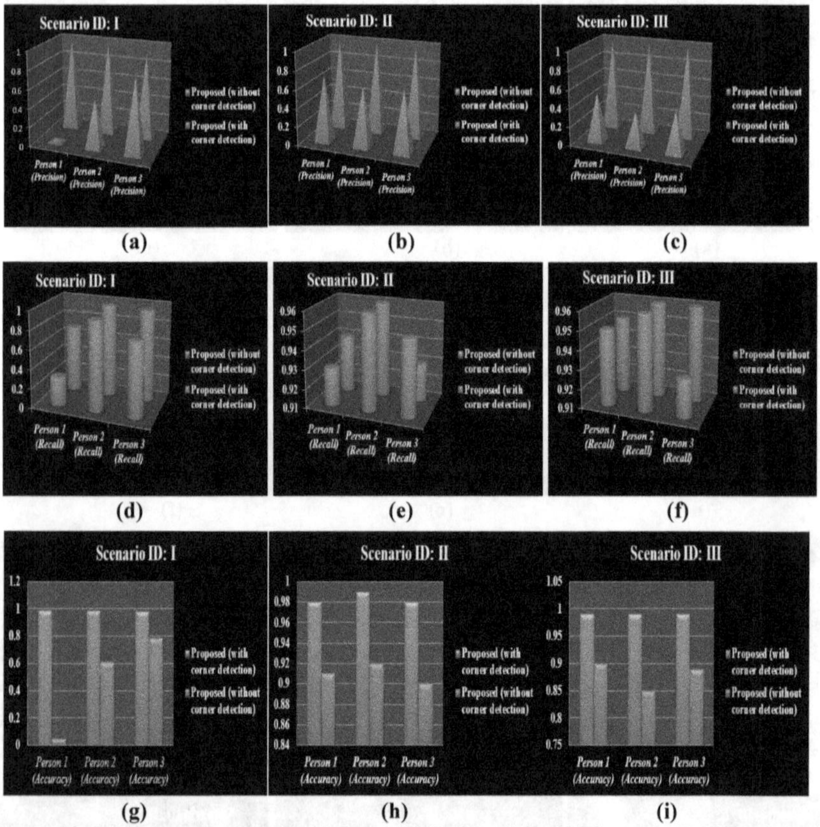

Figure 3.19: Interpretation graphs (a) Scenario ID I Precision (b) Scenario ID II Precision (c) Scenario ID III Precision (d) Scenario ID I Recall (e) Scenario ID II Recall (f) Scenario ID III Recall (g) Scenario ID I Accuracy (h) Scenario ID II Accuracy (i) Scenario ID III Accuracy

Figure 3.16 shows the error between the actual and the anticipated trajectory for all humans in different scenarios. Here the proposed model renders many fewer errors as compared to other methods. F-score is represented in Figure 3.17. Multiple confusion parameters can be visualised in Figure 3.18 where the proposed model gives a good confidence score. Precision, recall, and accuracy has been represented in Figure 3.19 for different scenarios and it is compared with a corner non-corner detection heuristic.

3.8.3.2 Visualisation of the anticipated trajectory and its comparison with ground truth

This thesis applied the proposed 3D particle model in different scenarios and each scenario elaborates multi-behavioural human tracking. Here the humans sometimes leave and re-join the group, swap their positions, and also overtake each other which can be visualised in the given figures. The tracker is deleted when the human leaves the group and is automatically re-initialised when the humans join the group to follow the robot. Figure 3.20, Figure 3.28, and Figure 3.36 are the graphs for the anticipated trajectory of the humans based on three different scenarios respectively using the Multi-behavioural 3D Particle Social Force Model.

The errors may seem larger than conventionally quoted in the literature for a tracker with a mobile observer model. However, three differences need to be noted in the proposed approach. First that the observer (robot) constant loses the visibility of the human. In the conventional approaches, the errors exponentially rise with time under such settings. The proposed approach reduces the errors using the social model. Second, that if the person re-joins the group, only a monocular camera is available, and the position needs to be guessed. The visibility of the person upon re-joining shall be an extremity in the image, and the cameras are very sensitive in that region. Therefore, the re-initialisation has a very large error that the tracker must overcome. Finally, the humans teleoperators were asked to act erratically in the control of the robot. There may have been reasonably long durations of time when the robot was not visible to the camera and the human teleoperator was not aware about it. In such cases, the teleoperator may have also been using erratic behaviours that do not display the social norms. These limitations exist in all algorithms, however, only the proposed algorithm can address these limitations sufficiently well.

The human speeds and orientation are also considered to best fit the proposed model which is represented in Figure 3.21, Figure 3.29, and Figure 3.37 for the different scenarios. Multiple attraction and repulsive force are utilised to model this 3D social force tracking module and are shown in the given Figure 3.22, Figure 3.30, and Figure 3.38. Tracking results based on random particle filter are drawn and explained in Figure 3.23, Figure 3.31, and Figure 3.39 which do not generate accurate human trajectory. Furthermore the 'Tracking in Image' method is not able to track the humans in these scenarios and predicts very poor trajectories as soon as the robot is out of the field of vision, which constitutes a large part of the scenarios, which can be visualised in Figure 3.24, Figure 3.32, and Figure 3.40. Additionally, the thesis has also performed the Kalman Filter approach for the trajectory prediction as shown in Figure 3.25, Figure 3.33, and Figure 3.41

and this method is not able to anticipate the positions of the humans where there is no visibility or at the corners.

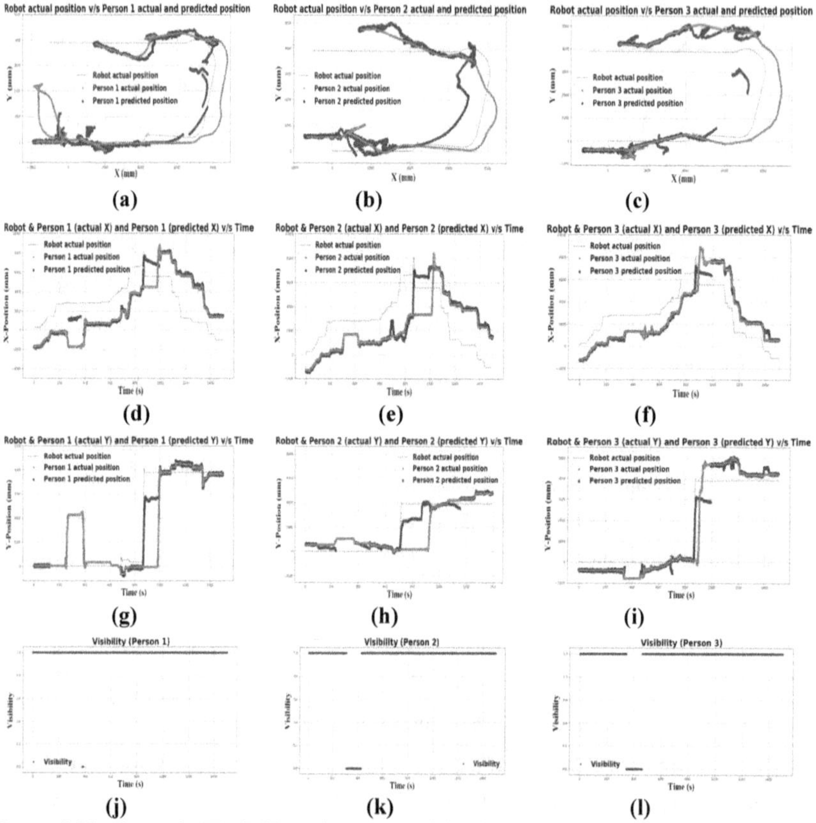

Figure 3.20: Scenario ID: I 3D trajectory anticipation on a different plane using a Multi-behavioural 3D Particle Social Force Model (a) Robot trajectory v/s human 1 actual v/s anticipated trajectory (b) Robot trajectory v/s human 2 actual v/s anticipated trajectory (c) Robot trajectory v/s human 3 actual v/s anticipated trajectory (d) Robot trajectory v/s human 1 actual v/s anticipated trajectory on X-plane (d) Robot trajectory v/s human 2 actual v/s anticipated trajectory on X-plane (f) Robot trajectory v/s human 3 actual v/s anticipated trajectory on X-plane (g) Robot trajectory v/s human 1 actual v/s anticipated trajectory on Y-plane (h) Robot trajectory v/s human 2 actual v/s anticipated trajectory on Y-plane (i) Robot trajectory v/s human 3 actual v/s anticipated trajectory on Y-plane (j) Visibility graph of human 1 (k) Visibility graph of human 2 (l) Visibility graph of human 3

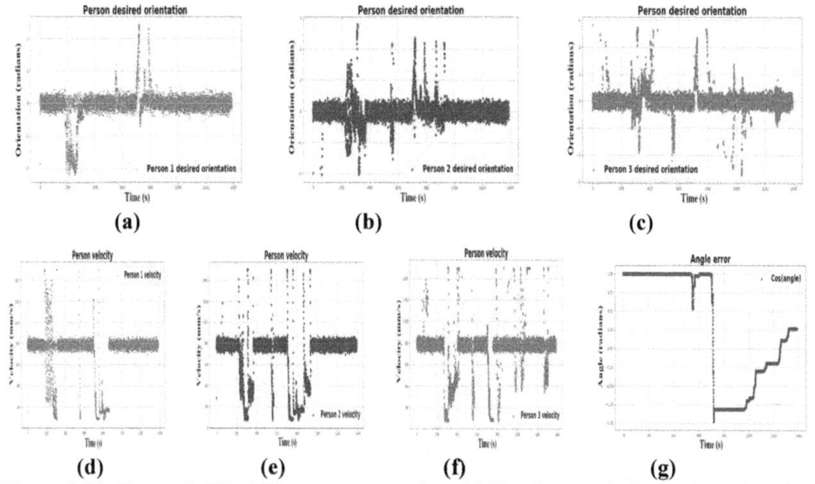

(a) (b) (c)

(d) (e) (f) (g)

Figure 3.21: Scenario ID: I Humans constraints (a) First human desired orientation (b) Second human desired orientation (c) Third human desired orientation (d) Velocity of first person (e) Velocity of second person (f) Velocity of third person (g) Corner angle error

To figure out the performance of the deep learning approach, this thesis applied an LSTM method with re-initialisation and without re-initialisation and the outcomes of LSTM with re-initialisation is visualised in Figure 3.26, Figure 3.34, and Figure 3.42 whereas the outcome of LSTM without re-initialisation is visualised in Figure 3.27, Figure 3.35, and Figure 3.43. The figures show that the four methods; random particle filter, tracking in the image, Kalman Filter, LSTM with re-initialisation, and LSTM without re-initialisation are not robust enough to handle the corner problem and human tracking at the corner is not possible by these approaches. The proposed model is robust enough to handle these situations, can easily detect when the robot is circumventing a corner leading to no visibility and performs tracking well.

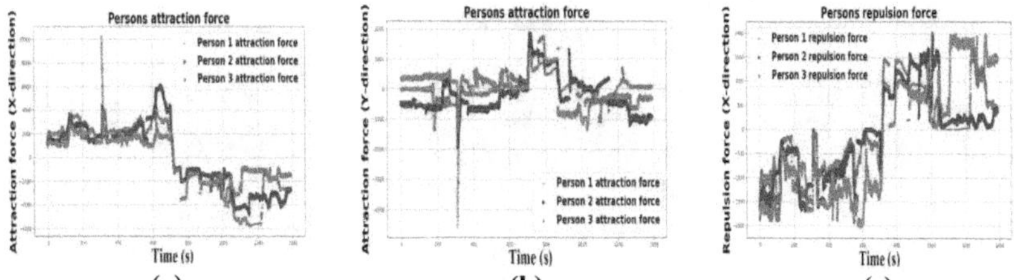

(a) (b) (c)

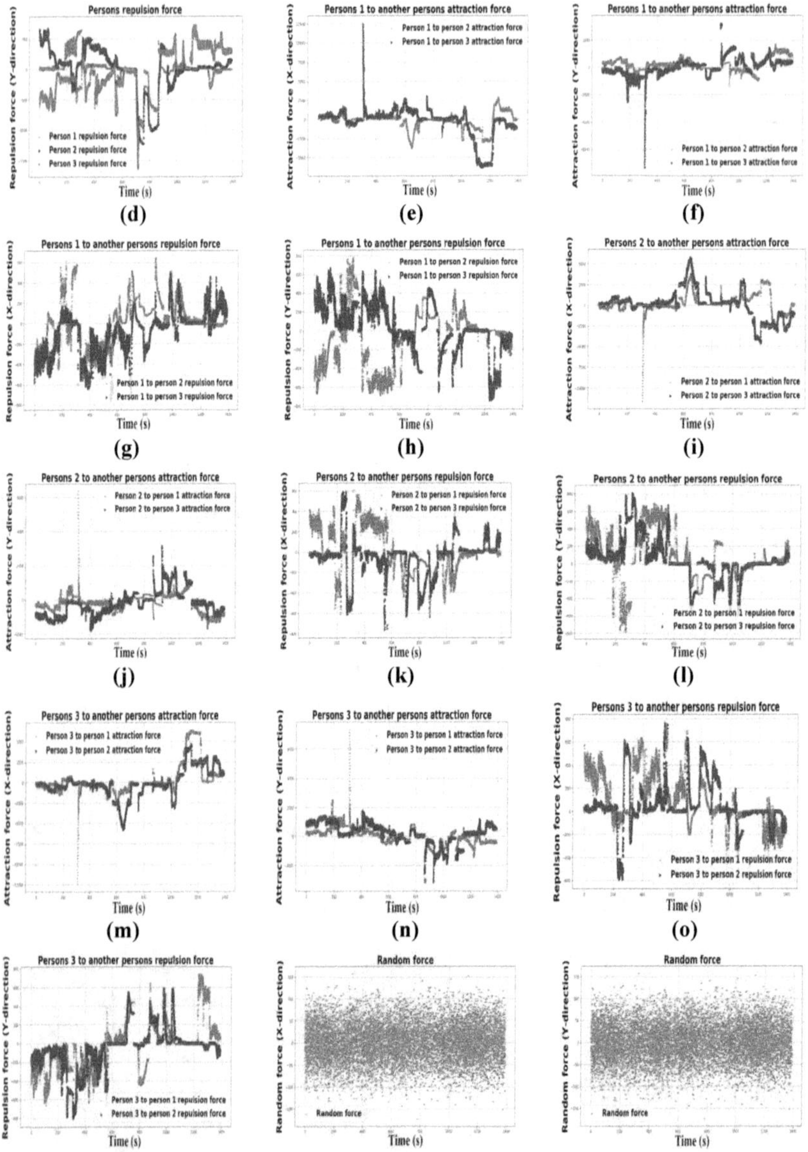

Persons repulsion force
Repulsion force (Y-direction)
Person 1 repulsion force
Person 2 repulsion force
Person 3 repulsion force
Time (s)
(d)
Persons 1 to another persons attraction force
Attraction force (X-direction)
Person 1 to person 2 attraction force
Person 1 to person 3 attraction force
Time (s)
(e)
Persons 1 to another persons attraction force
Attraction force (Y-direction)
Person 1 to person 2 attraction force
Person 1 to person 3 attraction force
Time (s)
(f)
Persons 1 to another persons repulsion force
Repulsion force (X-direction)
Person 1 to person 2 repulsion force
Person 1 to person 3 repulsion force
Time (s)
(g)
Persons 1 to another persons repulsion force
Repulsion force (Y-direction)
Person 1 to person 2 repulsion force
Person 1 to person 3 repulsion force
Time (s)
(h)
Persons 2 to another persons attraction force
Attraction force (X-direction)
Person 2 to person 1 attraction force
Person 2 to person 3 attraction force
Time (s)
(i)
Persons 2 to another persons attraction force
Attraction force (Y-direction)
Person 2 to person 1 attraction force
Person 2 to person 3 attraction force
Time (s)
(j)
Persons 2 to another persons repulsion force
Repulsion force (X-direction)
Person 2 to person 1 repulsion force
Person 2 to person 3 repulsion force
Time (s)
(k)
Persons 2 to another persons repulsion force
Repulsion force (Y-direction)
Person 2 to person 1 repulsion force
Person 2 to person 3 repulsion force
Time (s)
(l)
Persons 3 to another persons attraction force
Attraction force (X-direction)
Person 3 to person 1 attraction force
Person 3 to person 2 attraction force
Time (s)
(m)
Persons 3 to another persons attraction force
Attraction force (Y-direction)
Person 3 to person 1 attraction force
Person 3 to person 2 attraction force
Time (s)
(n)
Persons 3 to another persons repulsion force
Repulsion force (X-direction)
Person 3 to person 1 repulsion force
Person 3 to person 2 repulsion force
Time (s)
(o)
Persons 3 to another persons repulsion force
Repulsion force (Y-direction)
Person 3 to person 1 repulsion force
Person 3 to person 2 repulsion force
Time (s)
(p)
Random force
Random force (X-direction)
Random force
Time (s)
(q)
Random force
Random force (Y-direction)
Random force
Time (s)
(r)

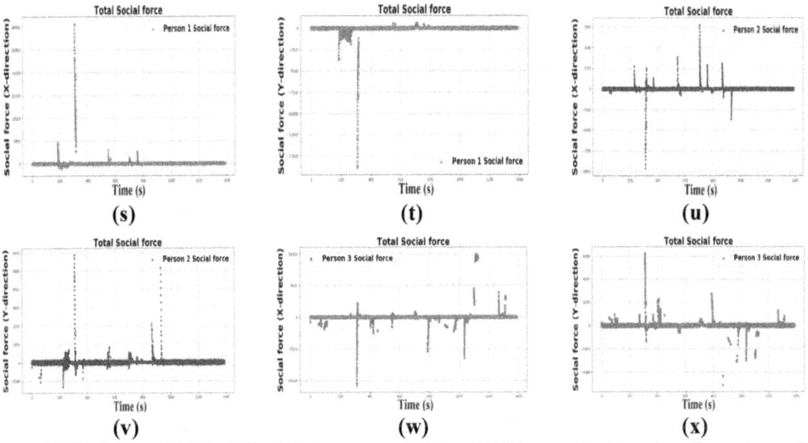

Figure 3.22: Scenario ID: I Social forces modelling (a) Humans' attraction force towards the robot in the X-direction (b) Humans' attraction force towards the robot in the Y-direction (c) Humans' repulsive force towards the robot in the X-direction (d) Humans' repulsive force towards the robot in the Y-direction (e) Person 1 to another person's attraction force in the X-direction (f) Person 1 to another person's attraction force in the Y-direction (g) Person 1 to another person's repulsive force in the X-direction (h) Person 1 to another person's repulsive force in the Y-direction (i) Person 2 to another person's attraction force in the X-direction (j) Person 2 to another person's attraction force in the Y-direction (k) Person 2 to another person's repulsive force in the X-direction (l) Person 2 to another person's repulsive force in the Y-direction (m) Person 3 to another person's attraction force in the X-direction (n) Person 3 to another person's attraction force in the Y-direction (o) Person 3 to another person's repulsive force in the X-direction (p) Person 3 to another person's repulsive force in the Y-direction (q) Random force on X-plane (r) Random force on Y-plane (s) Total social force applied on human 1 in the X-direction (t) Total social force applied on human 1 in the Y-direction (u) Total social force applied on human 2 in the X-direction (v) Total social force applied on human 2 in the Y-direction (w) Total social force applied on human 3 in the X-direction (x) Total social force applied on human 3 in the Y-direction

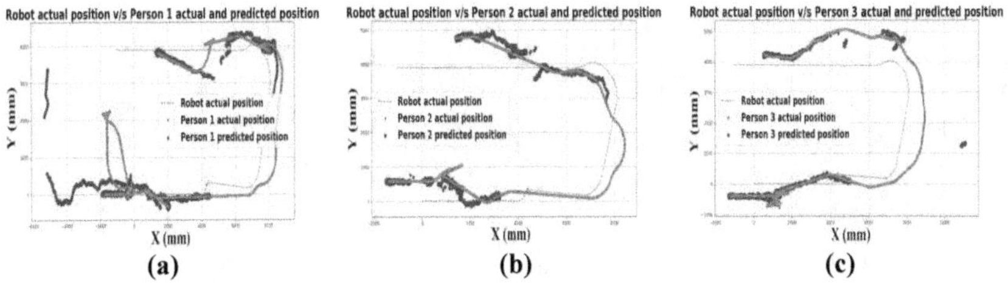

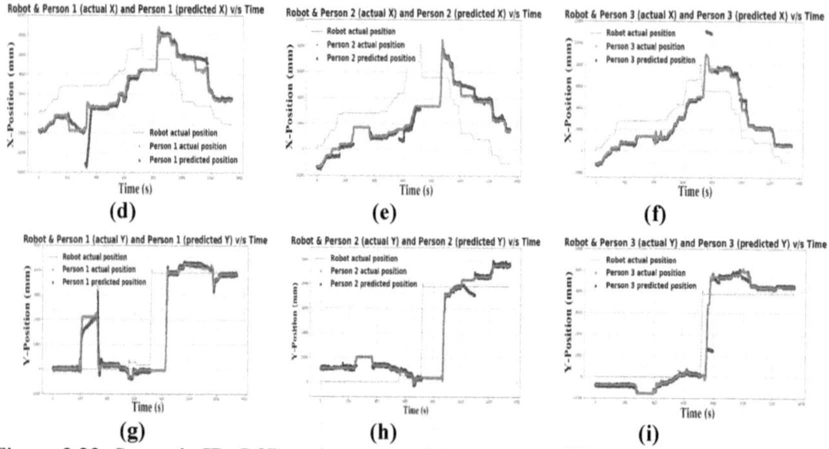

Figure 3.23: Scenario ID: I 3D trajectory anticipation on a different plane using a Random particle filter (a) Robot trajectory v/s human 1 actual v/s anticipated trajectory (b) Robot trajectory v/s human 2 actual v/s anticipated trajectory (c) Robot trajectory v/s human 3 actual v/s anticipated trajectory (d) Robot trajectory v/s human 1 actual v/s anticipated trajectory on X-plane (d) Robot trajectory v/s human 2 actual v/s anticipated trajectory on X-plane (f) Robot trajectory v/s human 3 actual v/s anticipated trajectory on X-plane (g) Robot trajectory v/s human 1 actual v/s anticipated trajectory on Y-plane (h) Robot trajectory v/s human 2 actual v/s anticipated trajectory on Y-plane (i) Robot trajectory v/s human 3 actual v/s anticipated trajectory on Y-plane

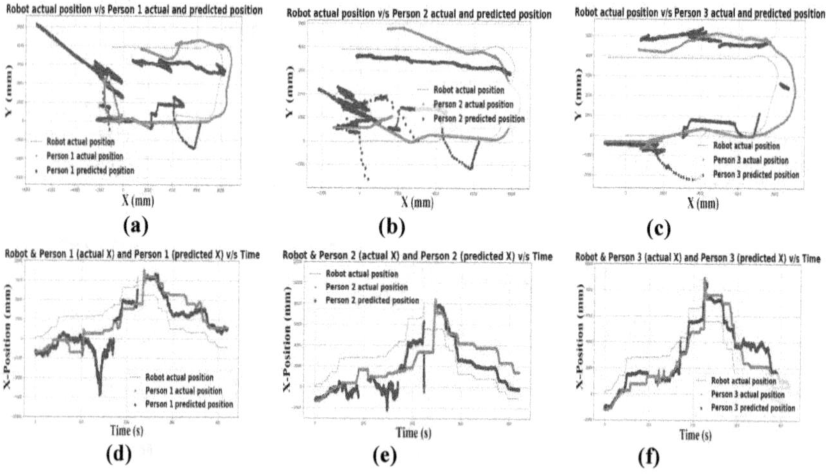

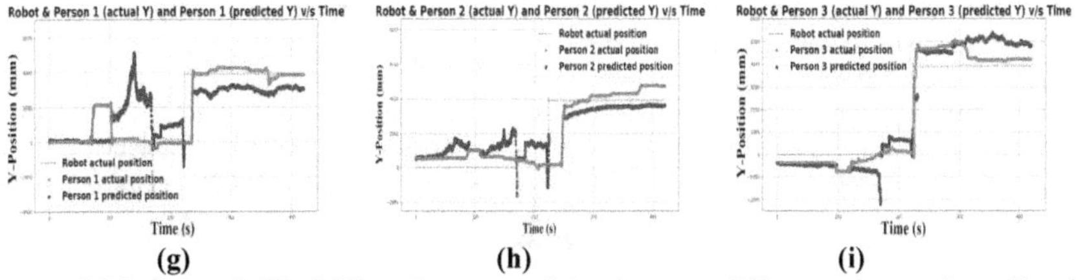

Figure 3.24: Scenario ID: I 3D trajectory anticipation on a different plane using a Tracking in image (a) Robot trajectory v/s human 1 actual v/s anticipated trajectory (b) Robot trajectory v/s human 2 actual v/s anticipated trajectory (c) Robot trajectory v/s human 3 actual v/s anticipated trajectory (d) Robot trajectory v/s human 1 actual v/s anticipated trajectory on X-plane (d) Robot trajectory v/s human 2 actual v/s anticipated trajectory on X-plane (f) Robot trajectory v/s human 3 actual v/s anticipated trajectory on X-plane (g) Robot trajectory v/s human 1 actual v/s anticipated trajectory on Y-plane (h) Robot trajectory v/s human 2 actual v/s anticipated trajectory on Y-plane (i) Robot trajectory v/s human 3 actual v/s anticipated trajectory on Y-plane

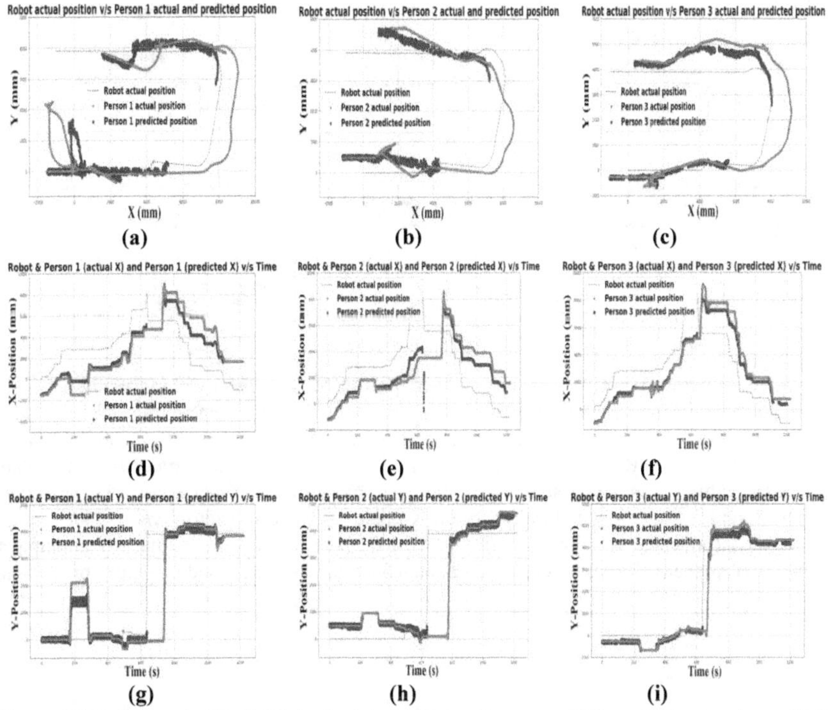

Figure 3.25: Scenario ID: I 3D trajectory anticipation on a different plane using a Kalman Filter (a) Robot trajectory v/s human 1 actual v/s anticipated trajectory (b) Robot trajectory v/s human 2 actual v/s anticipated trajectory (c) Robot trajectory v/s human 3 actual v/s anticipated

trajectory (d) Robot trajectory v/s human 1 actual v/s anticipated trajectory on X-plane (d) Robot trajectory v/s human 2 actual v/s anticipated trajectory on X-plane (f) Robot trajectory v/s human 3 actual v/s anticipated trajectory on X-plane (g) Robot trajectory v/s human 1 actual v/s anticipated trajectory on Y-plane (h) Robot trajectory v/s human 2 actual v/s anticipated trajectory on Y-plane (i) Robot trajectory v/s human 3 actual v/s anticipated trajectory on Y-plane

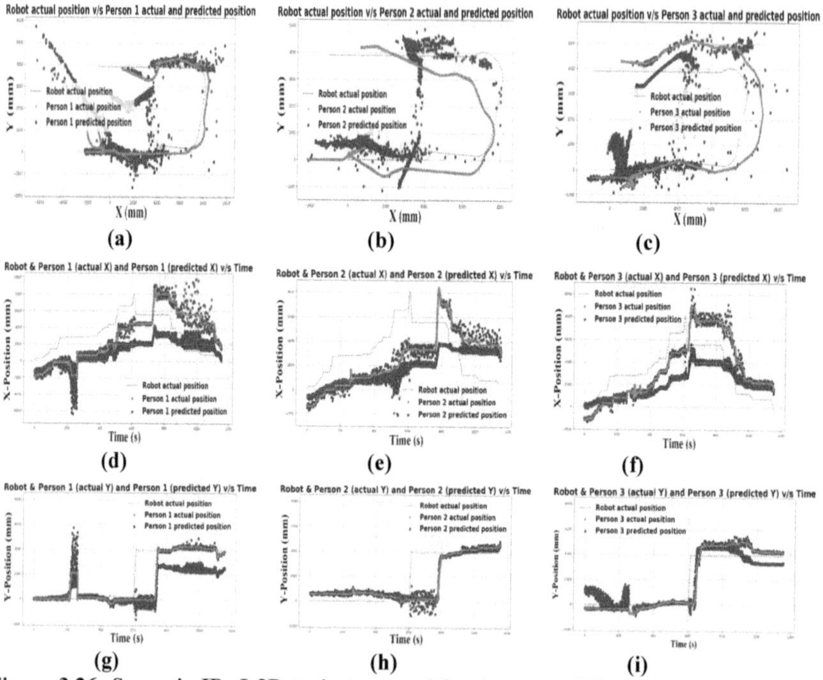

Figure 3.26: Scenario ID: I 3D trajectory anticipation on a different plane using a LSTM (with re-initialisation) (a) Robot trajectory v/s human 1 actual v/s anticipated trajectory (b) Robot trajectory v/s human 2 actual v/s anticipated trajectory (c) Robot trajectory v/s human 3 actual v/s anticipated trajectory (d) Robot trajectory v/s human 1 actual v/s anticipated trajectory on X-plane (d) Robot trajectory v/s human 2 actual v/s anticipated trajectory on X-plane (f) Robot trajectory v/s human 3 actual v/s anticipated trajectory on X-plane (g) Robot trajectory v/s human 1 actual v/s anticipated trajectory on Y-plane (h) Robot trajectory v/s human 2 actual v/s anticipated trajectory on Y-plane (i) Robot trajectory v/s human 3 actual v/s anticipated trajectory on Y-plane

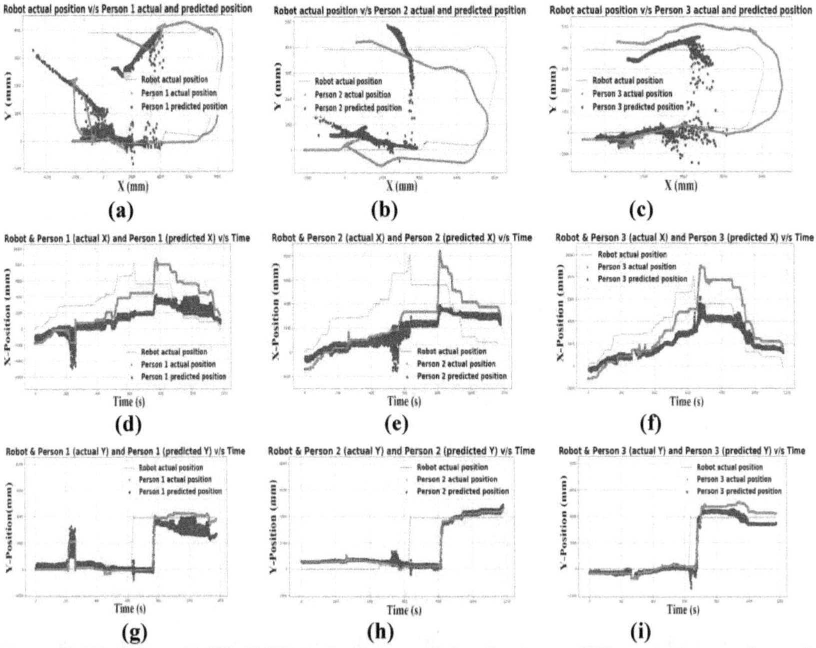

Figure 3.27: Scenario ID: I 3D trajectory anticipation on a different plane using a LSTM (without re-initialisation) (a) Robot trajectory v/s human 1 actual v/s anticipated trajectory (b) Robot trajectory v/s human 2 actual v/s anticipated trajectory (c) Robot trajectory v/s human 3 actual v/s anticipated trajectory (d) Robot trajectory v/s human 1 actual v/s anticipated trajectory on X-plane (d) Robot trajectory v/s human 2 actual v/s anticipated trajectory on X-plane (f) Robot trajectory v/s human 3 actual v/s anticipated trajectory on X-plane (g) Robot trajectory v/s human 1 actual v/s anticipated trajectory on Y-plane (h) Robot trajectory v/s human 2 actual v/s anticipated trajectory on Y-plane (i) Robot trajectory v/s human 3 actual v/s anticipated trajectory on Y-plane

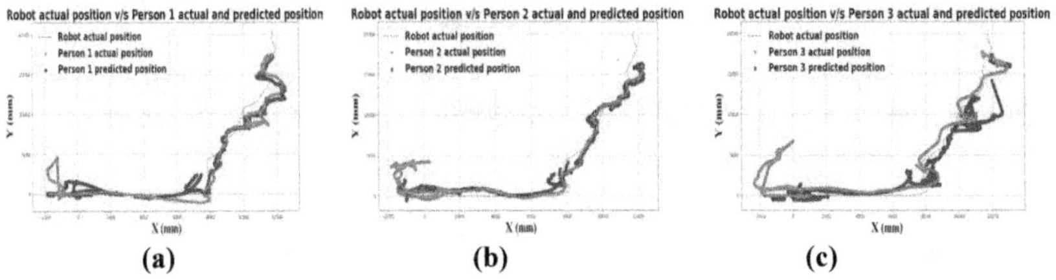

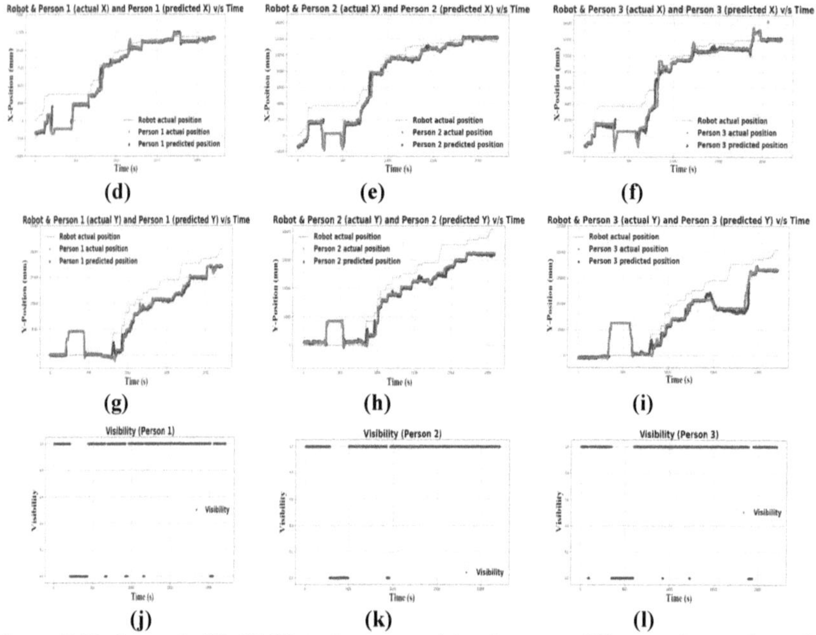

Figure 3.28: Scenario ID: II 3D trajectory anticipation on a different plane using a Multi-behavioural 3D Particle Social Force Model (a) Robot trajectory v/s human 1 actual v/s anticipated trajectory (b) Robot trajectory v/s human 2 actual v/s anticipated trajectory (c) Robot trajectory v/s human 3 actual v/s anticipated trajectory (d) Robot trajectory v/s human 1 actual v/s anticipated trajectory on X-plane (d) Robot trajectory v/s human 2 actual v/s anticipated trajectory on X-plane (f) Robot trajectory v/s human 3 actual v/s anticipated trajectory on X-plane (g) Robot trajectory v/s human 1 actual v/s anticipated trajectory on Y-plane (h) Robot trajectory v/s human 2 actual v/s anticipated trajectory on Y-plane (i) Robot trajectory v/s human 3 actual v/s anticipated trajectory on Y-plane (j) Visibility graph of human 1 (k) Visibility graph of human 2 (l) Visibility graph of human 3

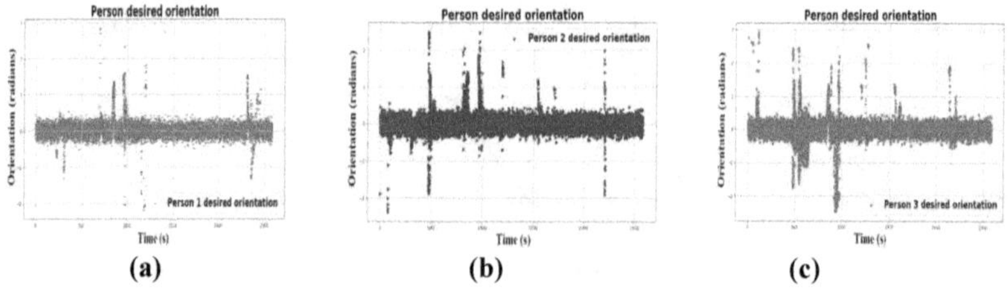

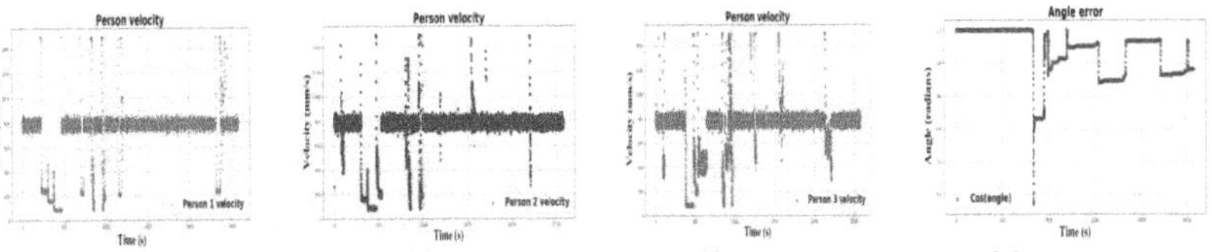

(d) (e) (f) (g)

Figure 3.29: Scenario ID: II Humans constraints (a) First human desired orientation (b) Second human desired orientation (c) Third human desired orientation (d) Velocity of first person (e) Velocity of second person (f) Velocity of third person (g) Corner angle error

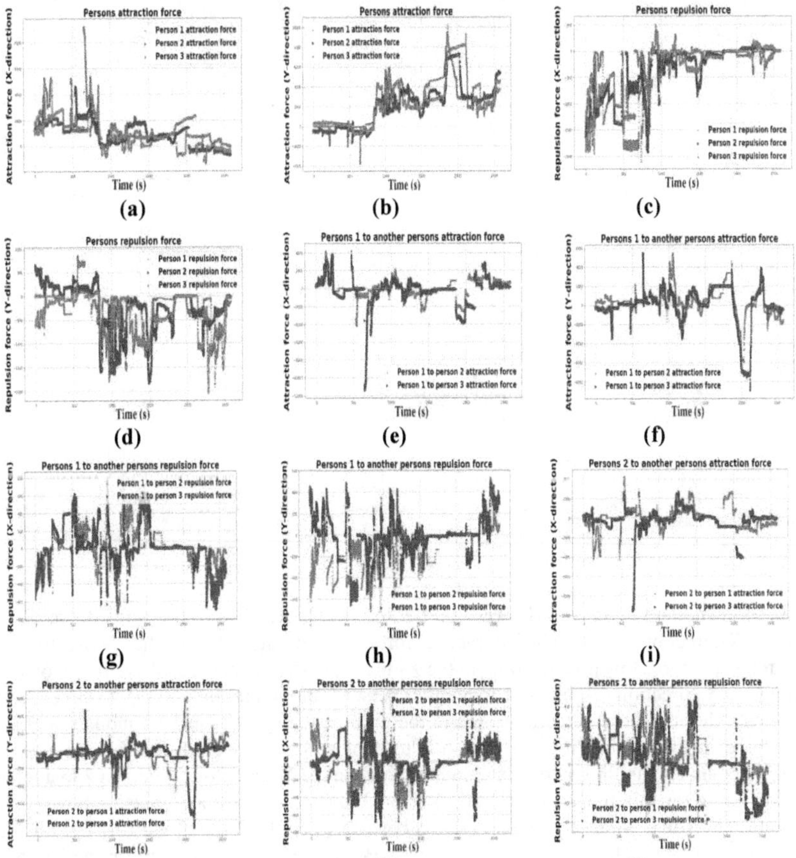

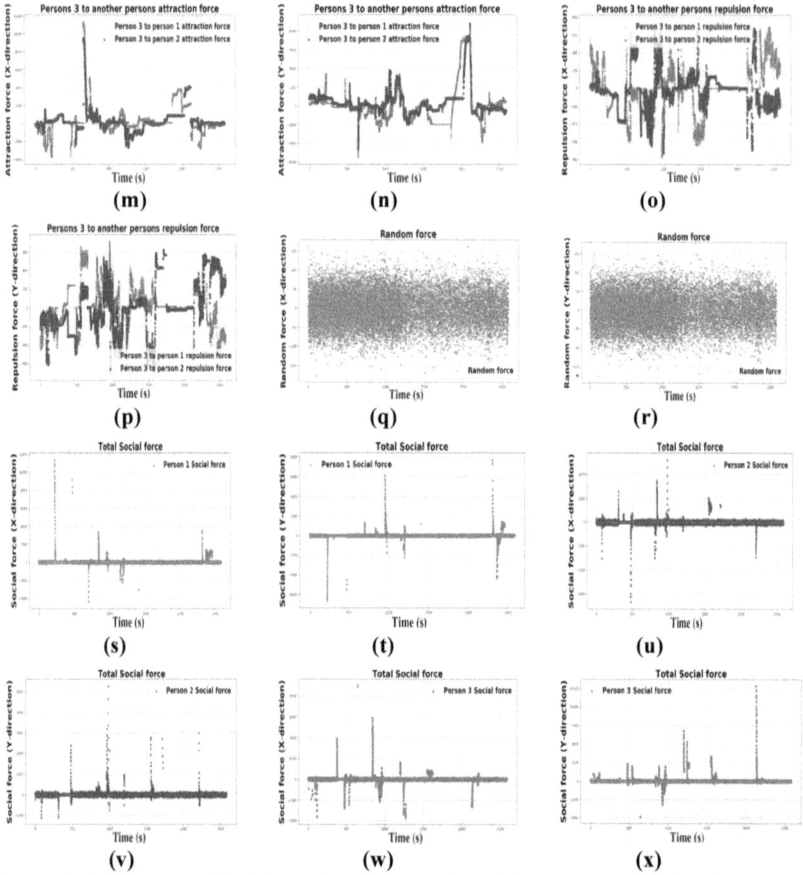

Figure 3.30: Scenario ID: II Social forces modelling (a) Humans' attraction force towards the robot in the X-direction (b) Humans' attraction force towards the robot in the Y-direction (c) Humans' repulsive force towards the robot in the X-direction (d) Humans' repulsive force towards the robot in the Y-direction (e) Person 1 to another person's attraction force in the X-direction (f) Person 1 to another person's attraction force in the Y-direction (g) Person 1 to another person's repulsive force in the X-direction (h) Person 1 to another person's repulsive force in the Y-direction (i) Person 2 to another person's attraction force in the X-direction (j) Person 2 to another person's attraction force in the Y-direction (k) Person 2 to another person's repulsive force in the X-direction (l) Person 2 to another person's repulsive force in the Y-direction (m) Person 3 to another person's attraction force in the X-direction (n) Person 3 to another person's attraction force in the Y-direction (o) Person 3 to another person's repulsive force in the X-direction (p) Person 3 to another person's repulsive force in the Y-direction (q) Random force on X-plane (r) Random force on Y-plane (s) Total social force applied on human 1 in the X-direction (t) Total social force

applied on human 1 in the Y-direction (u) Total social force applied on human 2 in the X-direction (v) Total social force applied on human 2 in the Y-direction (w) Total social force applied on human 3 in the X-direction (x) Total social force applied on human 3 in the Y-direction

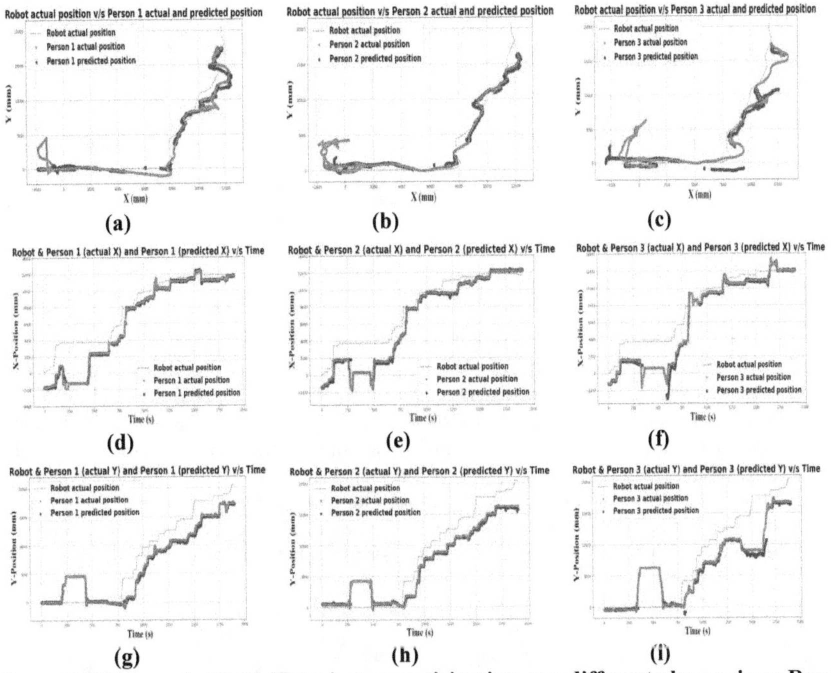

Figure 3.31: Scenario ID: II 3D trajectory anticipation on a different plane using a Random particle filter (a) Robot trajectory v/s human 1 actual v/s anticipated trajectory (b) Robot trajectory v/s human 2 actual v/s anticipated trajectory (c) Robot trajectory v/s human 3 actual v/s anticipated trajectory (d) Robot trajectory v/s human 1 actual v/s anticipated trajectory on X-plane (d) Robot trajectory v/s human 2 actual v/s anticipated trajectory on X-plane (f) Robot trajectory v/s human 3 actual v/s anticipated trajectory on X-plane (g) Robot trajectory v/s human 1 actual v/s anticipated trajectory on Y-plane (h) Robot trajectory v/s human 2 actual v/s anticipated trajectory on Y-plane (i) Robot trajectory v/s human 3 actual v/s anticipated trajectory on Y-plane

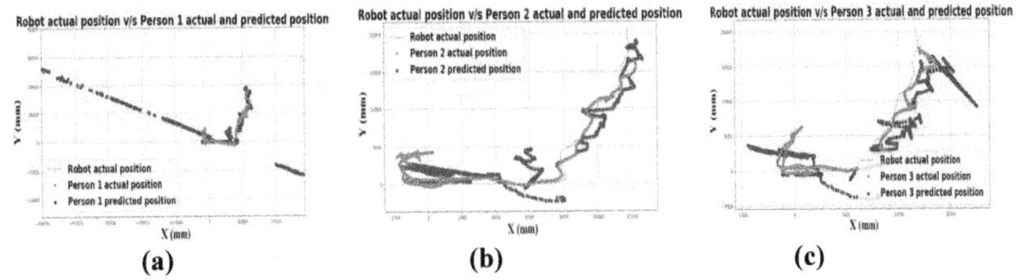

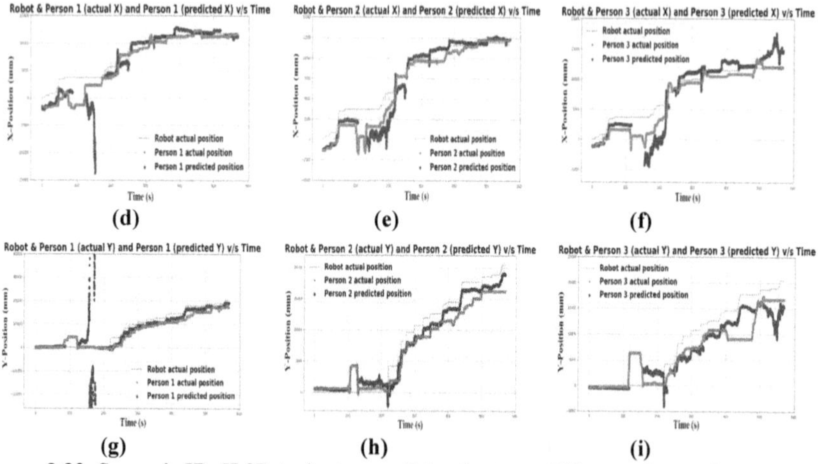

Figure 3.32: Scenario ID: II 3D trajectory anticipation on a different plane using a Tracking in image (a) Robot trajectory v/s human 1 actual v/s anticipated trajectory (b) Robot trajectory v/s human 2 actual v/s anticipated trajectory (c) Robot trajectory v/s human 3 actual v/s anticipated trajectory (d) Robot trajectory v/s human 1 actual v/s anticipated trajectory on X-plane (d) Robot trajectory v/s human 2 actual v/s anticipated trajectory on X-plane (f) Robot trajectory v/s human 3 actual v/s anticipated trajectory on X-plane (g) Robot trajectory v/s human 1 actual v/s anticipated trajectory on Y-plane (h) Robot trajectory v/s human 2 actual v/s anticipated trajectory on Y-plane (i) Robot trajectory v/s human 3 actual v/s anticipated trajectory on Y-plane

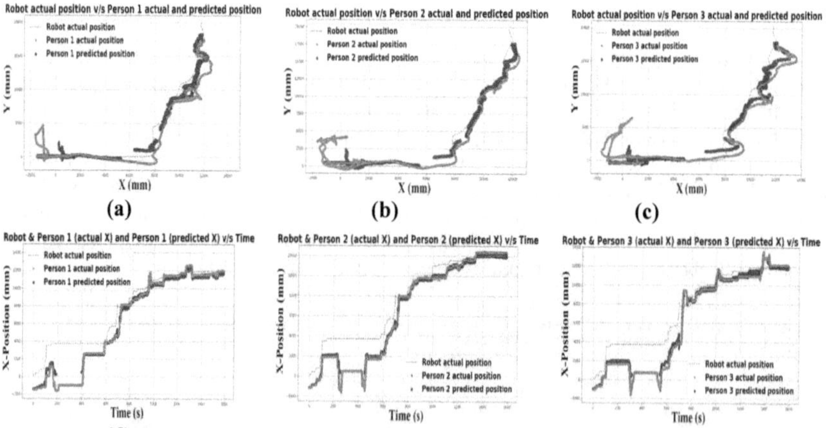

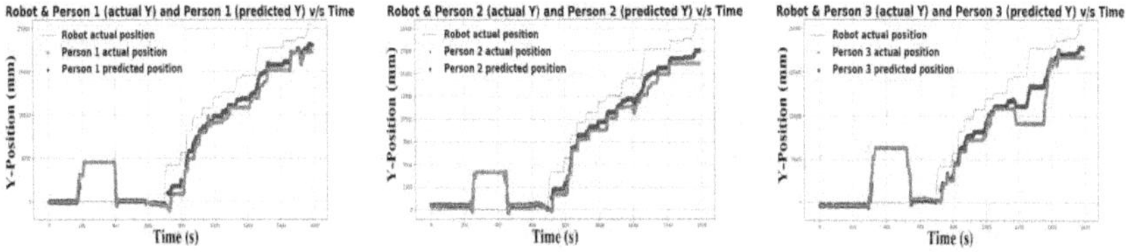

(g) (h) (i)

Figure 3.33: Scenario ID: II 3D trajectory anticipation on a different plane using a Kalman Filter (a) Robot trajectory v/s human 1 actual v/s anticipated trajectory (b) Robot trajectory v/s human 2 actual v/s anticipated trajectory (c) Robot trajectory v/s human 3 actual v/s anticipated trajectory (d) Robot trajectory v/s human 1 actual v/s anticipated trajectory on X-plane (d) Robot trajectory v/s human 2 actual v/s anticipated trajectory on X-plane (f) Robot trajectory v/s human 3 actual v/s anticipated trajectory on X-plane (g) Robot trajectory v/s human 1 actual v/s anticipated trajectory on Y-plane (h) Robot trajectory v/s human 2 actual v/s anticipated trajectory on Y-plane (i) Robot trajectory v/s human 3 actual v/s anticipated trajectory on Y-plane

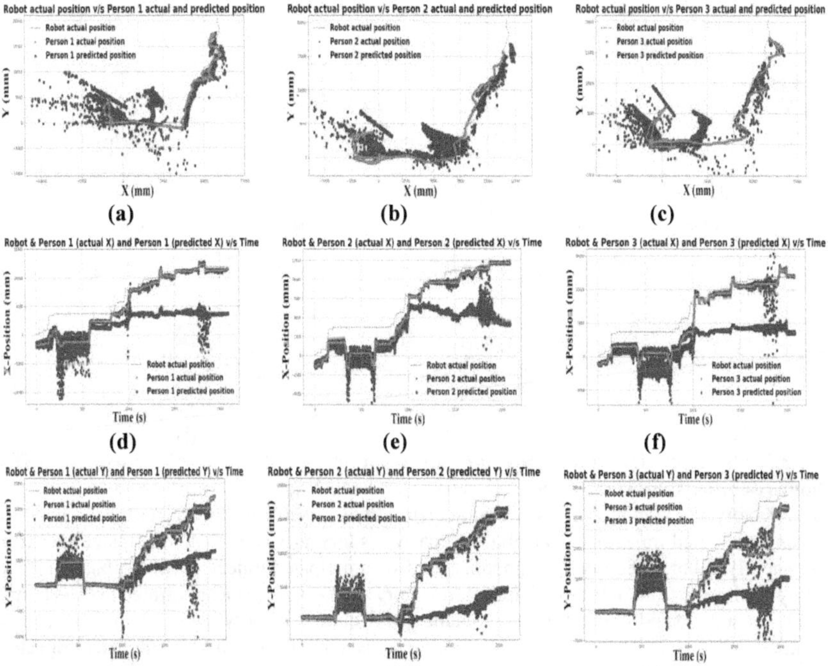

(g) (h) (i)

Figure 3.34: Scenario ID: II 3D trajectory anticipation on a different plane using a LSTM (with re-initialisation) (a) Robot trajectory v/s human 1 actual v/s anticipated trajectory (b) Robot trajectory v/s human 2 actual v/s anticipated trajectory (c) Robot trajectory v/s human 3 actual v/s anticipated trajectory (d) Robot trajectory v/s human 1 actual v/s anticipated trajectory on X-plane (d) Robot trajectory v/s human 2 actual v/s anticipated trajectory on X-plane (f) Robot trajectory v/s human 3 actual v/s anticipated trajectory on X-plane (g) Robot trajectory v/s human 1 actual v/s anticipated trajectory on Y-plane (h) Robot trajectory v/s human 2 actual v/s anticipated trajectory on Y-plane (i) Robot trajectory v/s human 3 actual v/s anticipated trajectory on Y-plane

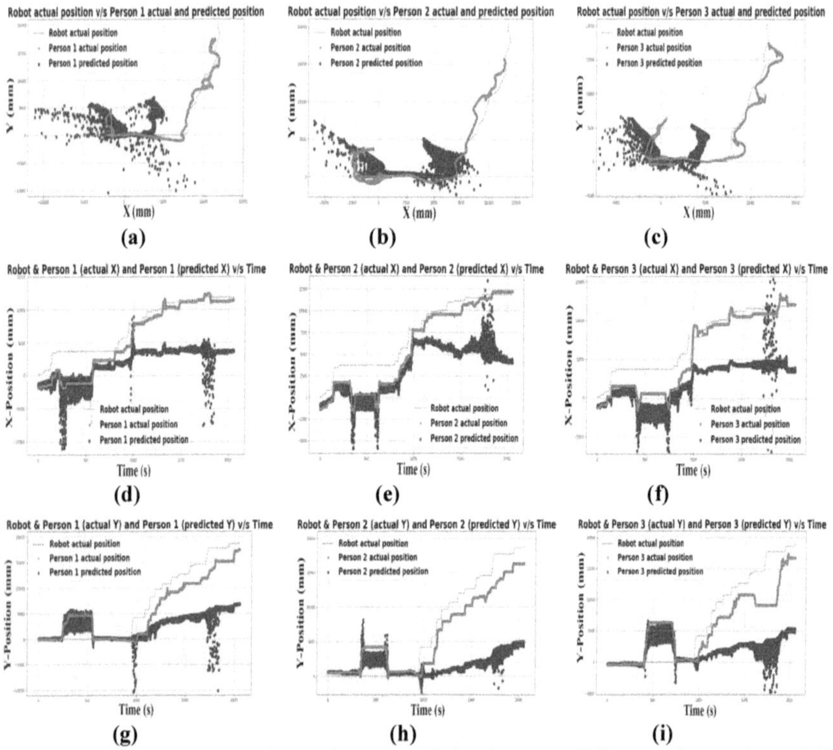

Figure 3.35: Scenario ID: II 3D trajectory anticipation on a different plane using a LSTM (without re-initialisation) (a) Robot trajectory v/s human 1 actual v/s anticipated trajectory (b) Robot trajectory v/s human 2 actual v/s anticipated trajectory (c) Robot trajectory v/s human 3 actual v/s anticipated trajectory (d) Robot trajectory v/s human 1 actual v/s anticipated trajectory on X-plane (d) Robot trajectory v/s human 2 actual v/s anticipated trajectory on X-plane (f) Robot trajectory v/s human 3 actual v/s anticipated trajectory on X-plane (g) Robot trajectory v/s human 1 actual v/s anticipated trajectory on Y-plane (h) Robot trajectory v/s human 2 actual v/s anticipated trajectory on Y-plane (i) Robot trajectory v/s human 3 actual v/s anticipated trajectory on Y-plane

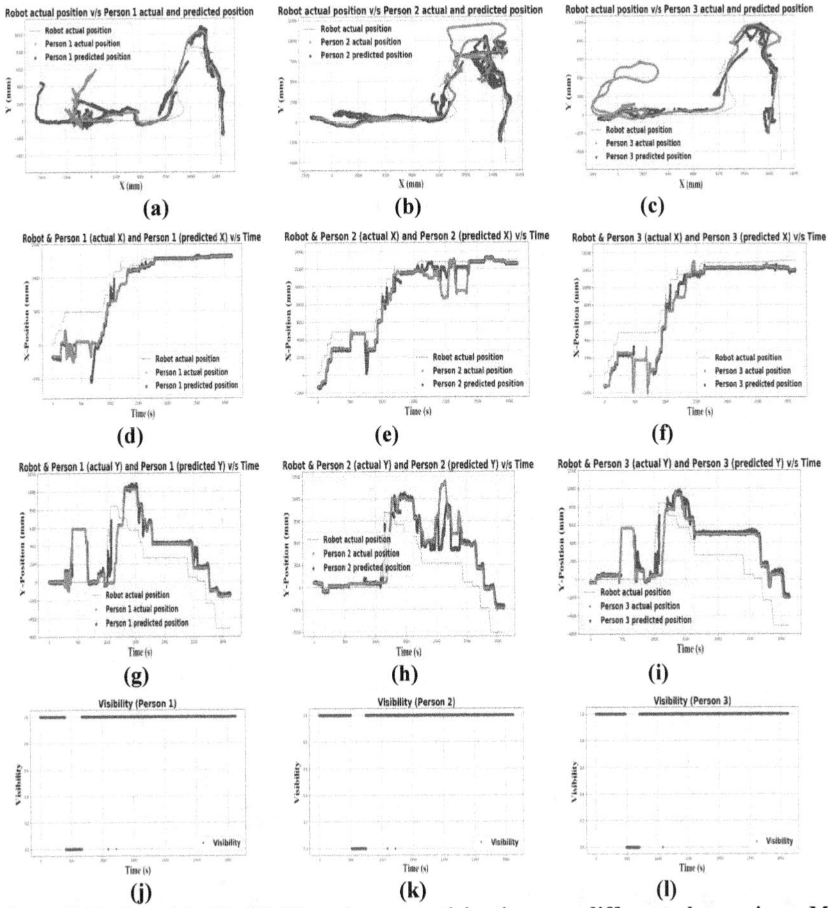

Figure 3.36: Scenario ID: III 3D trajectory anticipation on a different plane using a Multi-behavioural 3D Particle Social Force Model (a) Robot trajectory v/s human 1 actual v/s anticipated trajectory (b) Robot trajectory v/s human 2 actual v/s anticipated trajectory (c) Robot trajectory v/s human 3 actual v/s anticipated trajectory (d) Robot trajectory v/s human 1 actual v/s anticipated trajectory on X-plane (d) Robot trajectory v/s human 2 actual v/s anticipated trajectory on X-plane (f) Robot trajectory v/s human 3 actual v/s anticipated trajectory on X-plane (g) Robot trajectory v/s human 1 actual v/s anticipated trajectory on Y-plane (h) Robot trajectory v/s human 2 actual v/s anticipated trajectory on Y-plane (i) Robot trajectory v/s human 3 actual v/s anticipated trajectory on Y-plane (j) Visibility graph of human 1 (k) Visibility graph of human 2 (l) Visibility graph of human 3

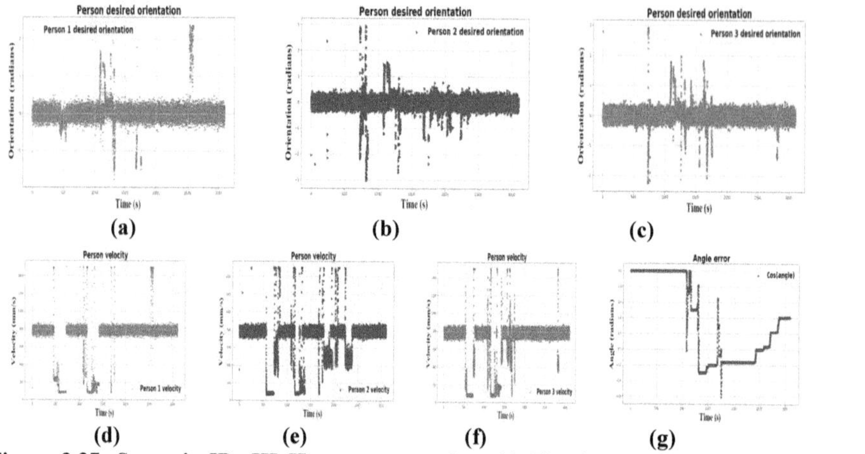

Figure 3.37: Scenario ID: III Humans constraints (a) First human desired orientation (b) Second human desired orientation (c) Third human desired orientation (d) Velocity of first person (e) Velocity of second person (f) Velocity of third person (g) Corner angle error

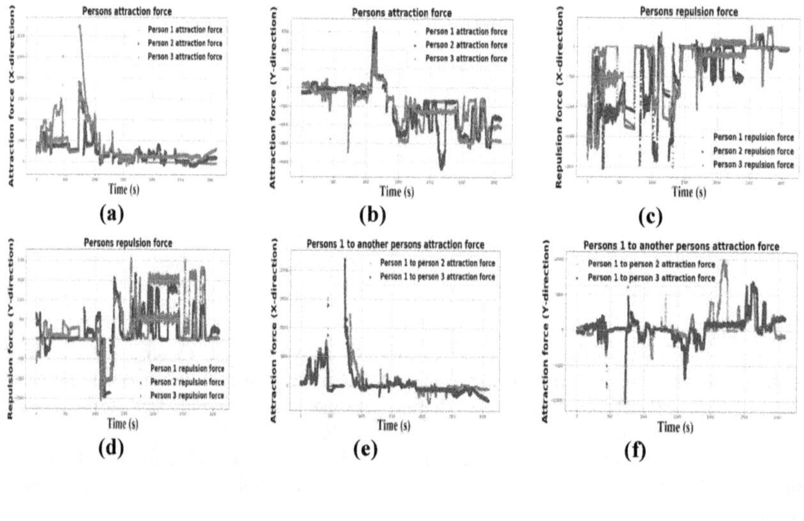

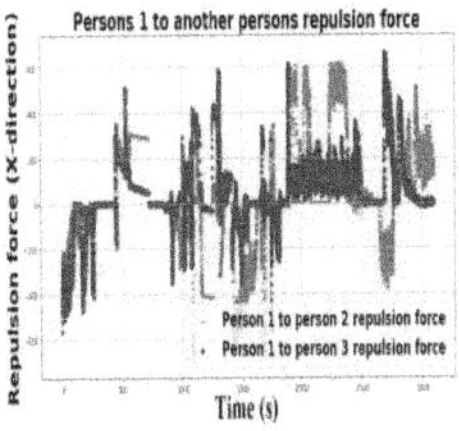

(g)

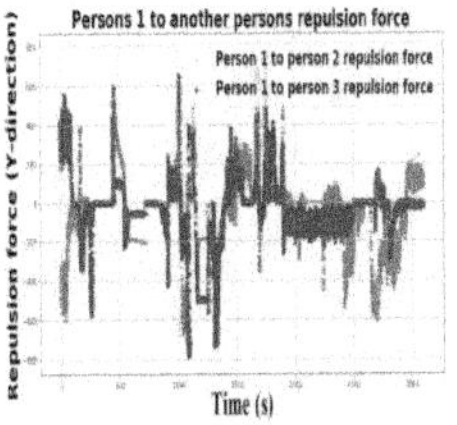

(h)

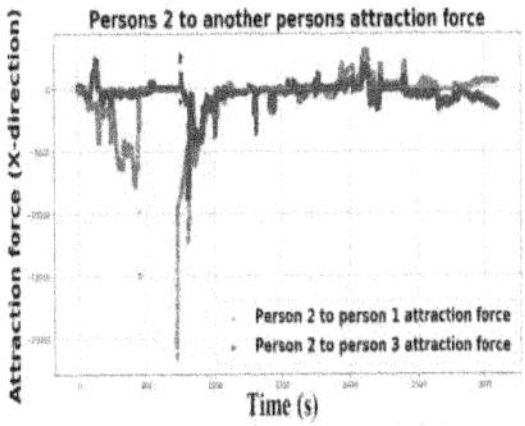

(i)

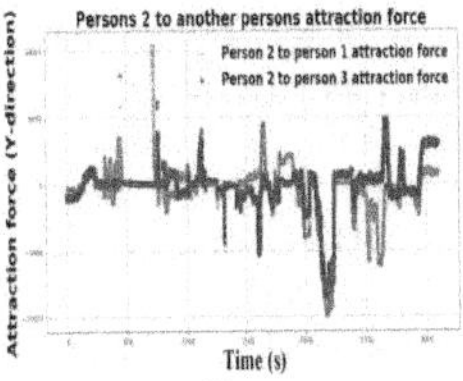

(j)

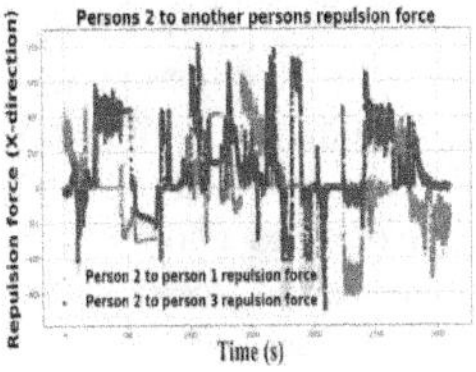

(k)

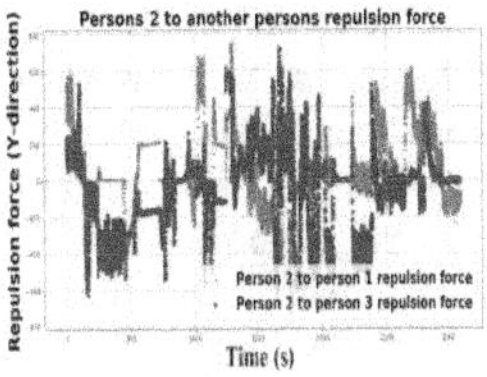

(l)

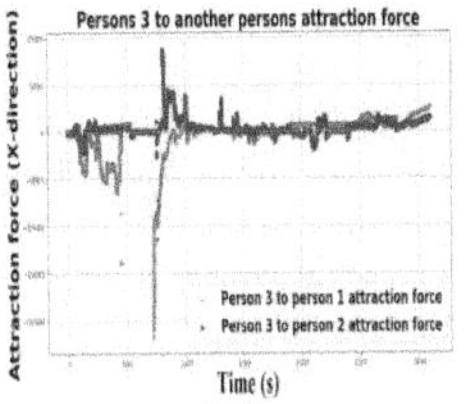

(m)

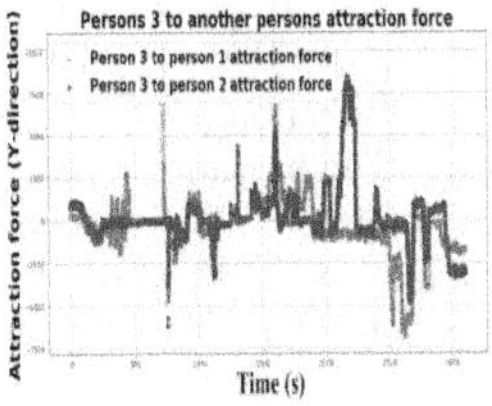

(n)

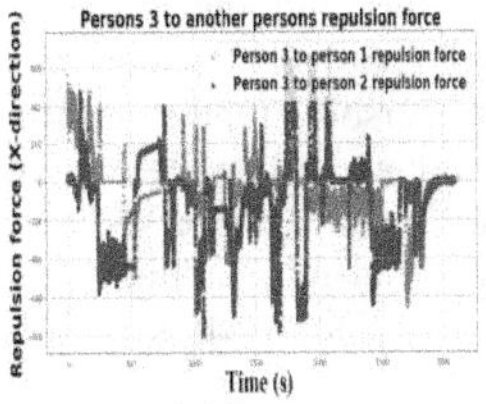

(o)

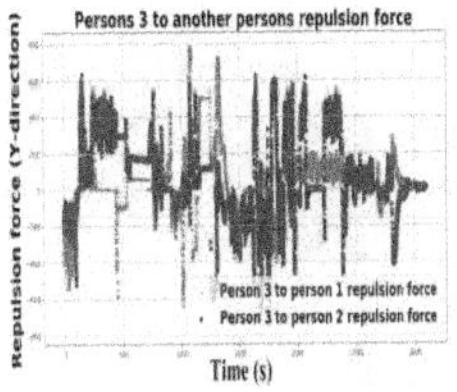

(p)

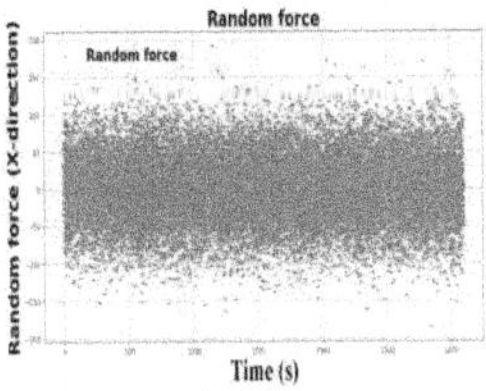

(q)

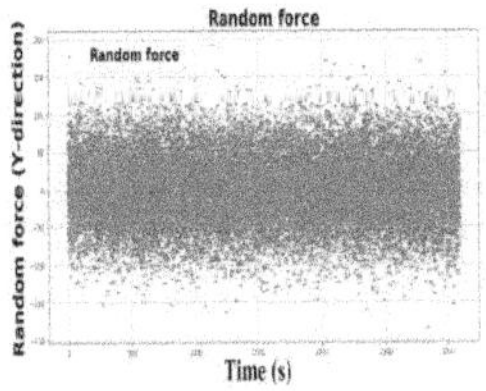

(r)

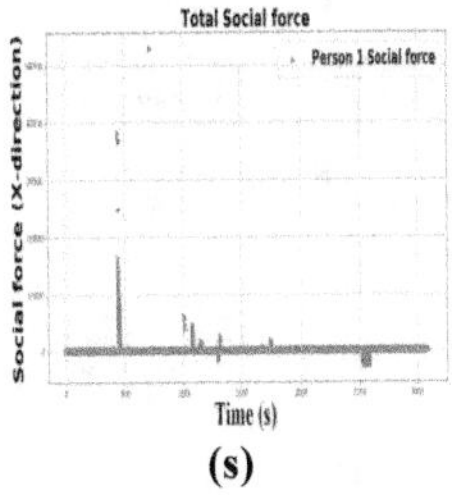

(s)

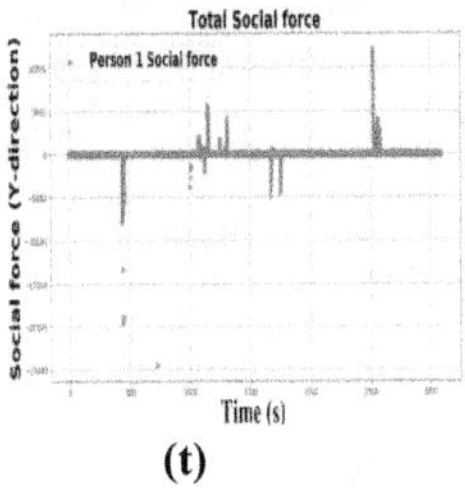

(t)

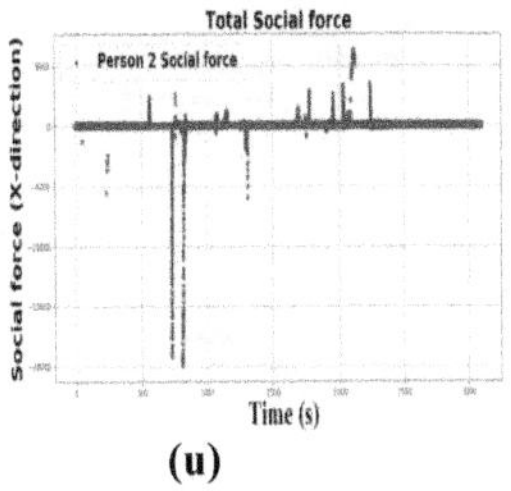

(u)

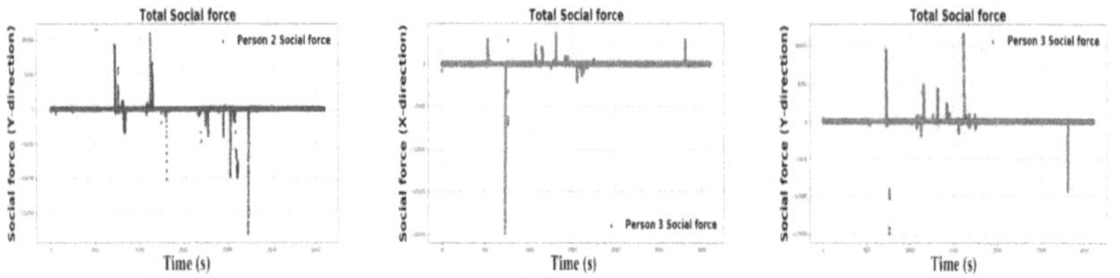

(v) **(w)** **(x)**

Figure 3.38: Scenario ID: III Social forces modelling (a) Humans' attraction force towards the robot in the X-direction (b) Humans' attraction force towards the robot in the Y-direction (c) Humans' repulsive force towards the robot in the X-direction (d) Humans' repulsive force towards the robot in the Y-direction (e) Person 1 to another person's attraction force in the X-direction (f) Person 1 to another person's attraction force in the Y-direction (g) Person 1 to another person's repulsive force in the X-direction (h) Person 1 to another person's repulsive force in the Y-direction (i) Person 2 to another person's attraction force in the X-direction (j) Person 2 to another person's attraction force in the Y-direction (k) Person 2 to another person's repulsive force in the X-direction (l) Person 2 to another person's repulsive force in the Y-direction (m) Person 3 to another person's attraction force in the X-direction (n) Person 3 to another person's attraction force in the Y-direction (o) Person 3 to another person's repulsive force in the X-direction (p) Person 3 to another person's repulsive force in the Y-direction (q) Random force on X-plane (r) Random force on Y-plane (s) Total social force applied on human 1 in the X-direction (t) Total social force applied on human 1 in the Y-direction (u) Total social force applied on human 2 in the X-direction (v) Total social force applied on human 2 in the Y-direction (w) Total social force applied on human 3 in the X-direction (x) Total social force applied on human 3 in the Y-direction

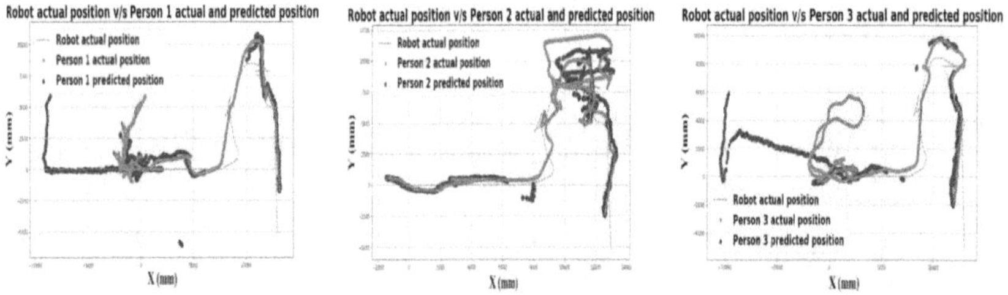

(a) **(b)** **(c)**

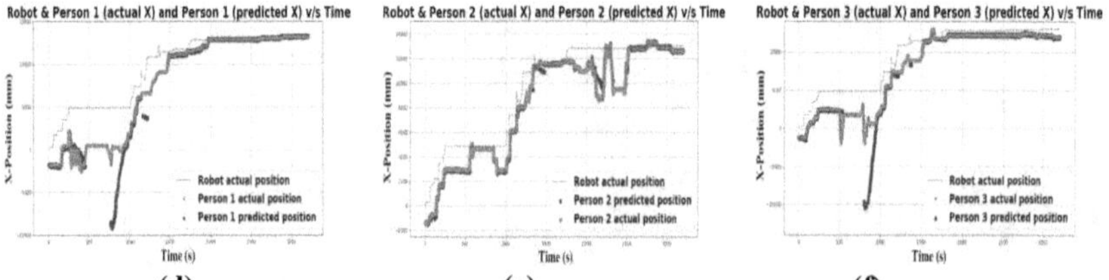

(d) **(e)** **(f)**

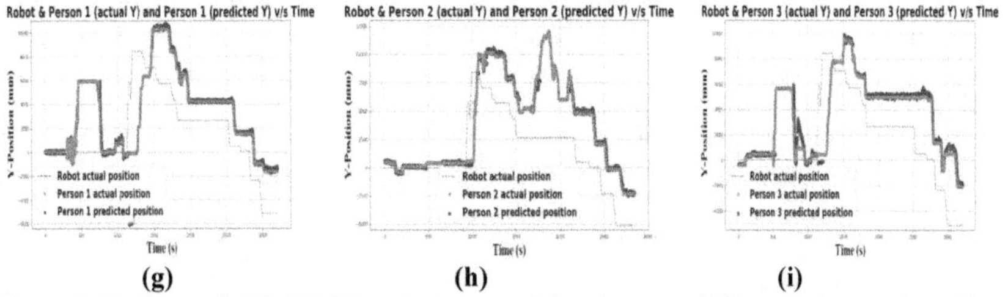

Figure 3.39: Scenario ID: III 3D trajectory anticipation on a different plane using a Random particle filter (a) Robot trajectory v/s human 1 actual v/s anticipated trajectory (b) Robot trajectory v/s human 2 actual v/s anticipated trajectory (c) Robot trajectory v/s human 3 actual v/s anticipated trajectory (d) Robot trajectory v/s human 1 actual v/s anticipated trajectory on X-plane (d) Robot trajectory v/s human 2 actual v/s anticipated trajectory on X-plane (f) Robot trajectory v/s human 3 actual v/s anticipated trajectory on X-plane (g) Robot trajectory v/s human 1 actual v/s anticipated trajectory on Y-plane (h) Robot trajectory v/s human 2 actual v/s anticipated trajectory on Y-plane (i) Robot trajectory v/s human 3 actual v/s anticipated trajectory on Y-plane

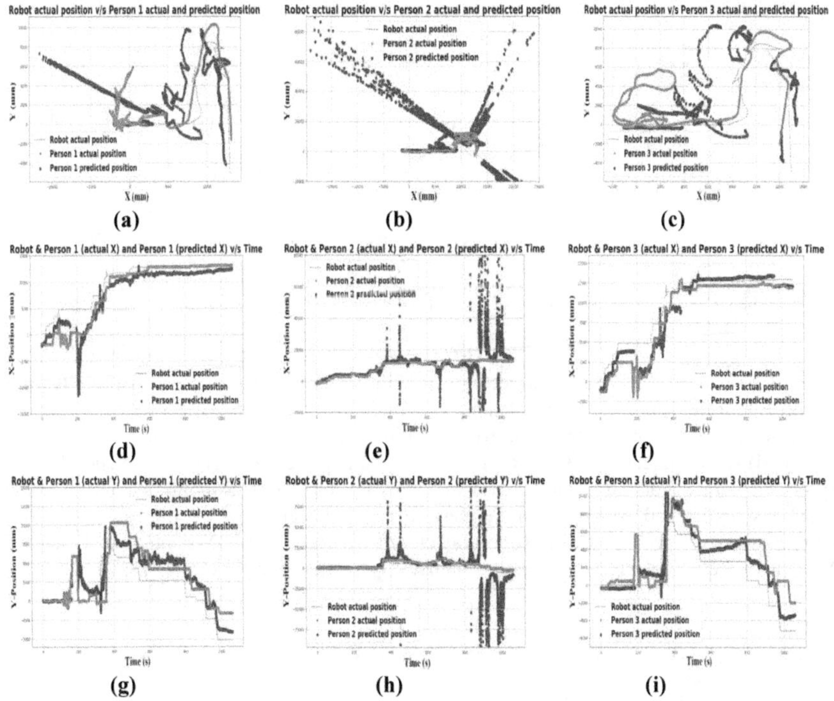

Figure 3.40: Scenario ID: III 3D trajectory anticipation on a different plane using a Tracking in image (a) Robot trajectory v/s human 1 actual v/s anticipated trajectory (b) Robot trajectory v/s

human 2 actual v/s anticipated trajectory (c) Robot trajectory v/s human 3 actual v/s anticipated trajectory (d) Robot trajectory v/s human 1 actual v/s anticipated trajectory on X-plane (d) Robot trajectory v/s human 2 actual v/s anticipated trajectory on X-plane (f) Robot trajectory v/s human 3 actual v/s anticipated trajectory on X-plane (g) Robot trajectory v/s human 1 actual v/s anticipated trajectory on Y-plane (h) Robot trajectory v/s human 2 actual v/s anticipated trajectory on Y-plane (i) Robot trajectory v/s human 3 actual v/s anticipated trajectory on Y-plane

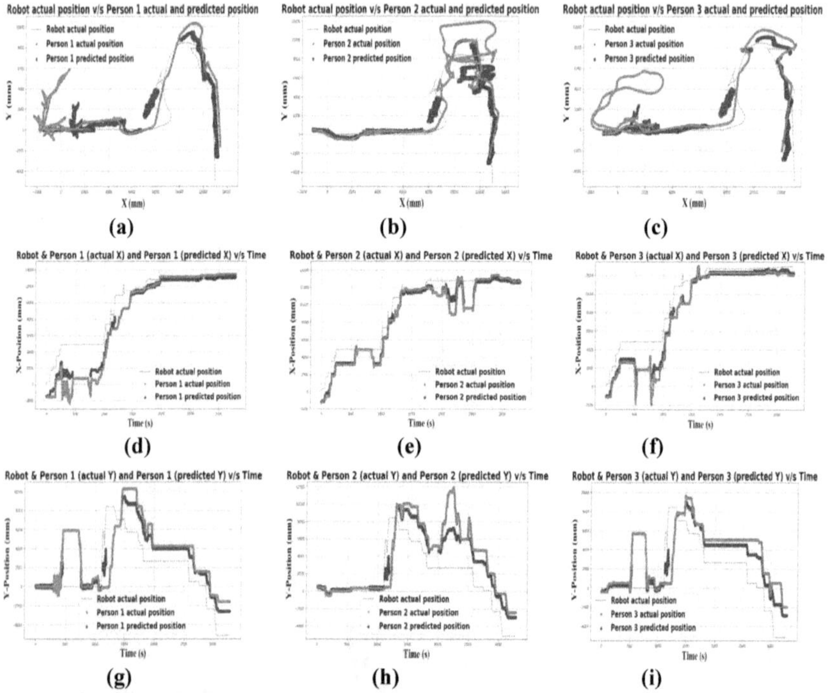

Figure 3.41: Scenario ID: III 3D trajectory anticipation on a different plane using a Kalman Filter (a) Robot trajectory v/s human 1 actual v/s anticipated trajectory (b) Robot trajectory v/s human 2 actual v/s anticipated trajectory (c) Robot trajectory v/s human 3 actual v/s anticipated trajectory (d) Robot trajectory v/s human 1 actual v/s anticipated trajectory on X-plane (d) Robot trajectory v/s human 2 actual v/s anticipated trajectory on X-plane (f) Robot trajectory v/s human 3 actual v/s anticipated trajectory on X-plane (g) Robot trajectory v/s human 1 actual v/s anticipated trajectory on Y-plane (h) Robot trajectory v/s human 2 actual v/s anticipated trajectory on Y-plane (i) Robot trajectory v/s human 3 actual v/s anticipated trajectory on Y-plane

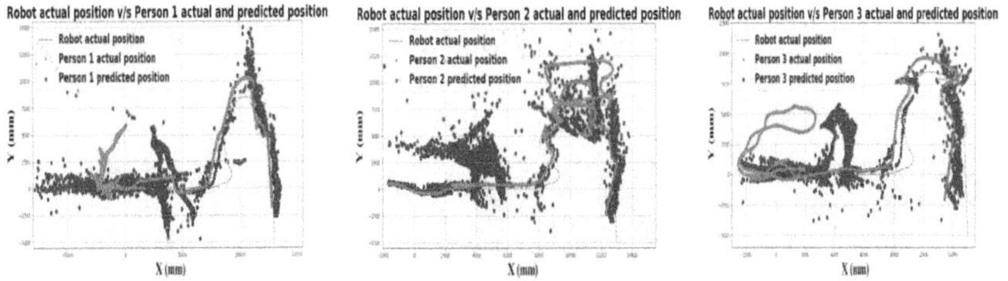

(a) **(b)** **(c)**

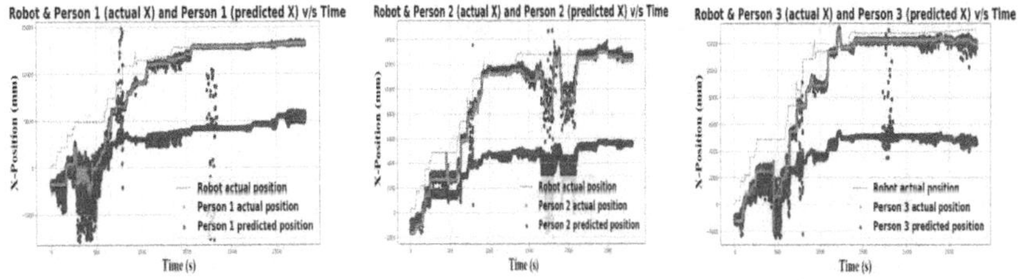

(d) **(e)** **(f)**

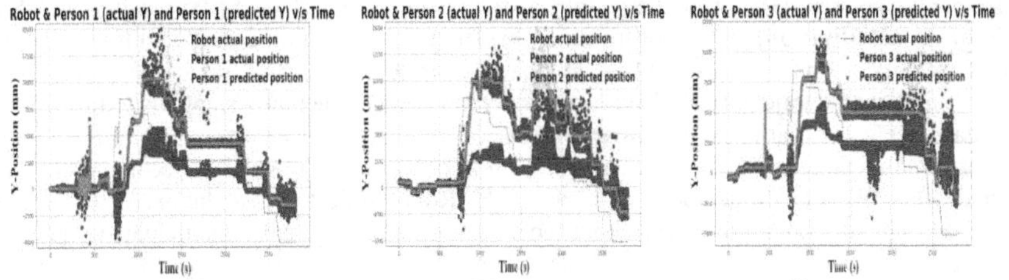

(g) **(h)** **(i)**

Figure 3.42: Scenario ID: III 3D trajectory anticipation on a different plane using a LSTM (with re-initialisation) (a) Robot trajectory v/s human 1 actual v/s anticipated trajectory (b) Robot trajectory v/s human 2 actual v/s anticipated trajectory (c) Robot trajectory v/s human 3 actual v/s anticipated trajectory (d) Robot trajectory v/s human 1 actual v/s anticipated trajectory on X-plane (d) Robot trajectory v/s human 2 actual v/s anticipated trajectory on X-plane (f) Robot trajectory v/s human 3 actual v/s anticipated trajectory on X-plane (g) Robot trajectory v/s human 1 actual v/s anticipated trajectory on Y-plane (h) Robot trajectory v/s human 2 actual v/s anticipated trajectory on Y-plane (i) Robot trajectory v/s human 3 actual v/s anticipated trajectory on Y-plane

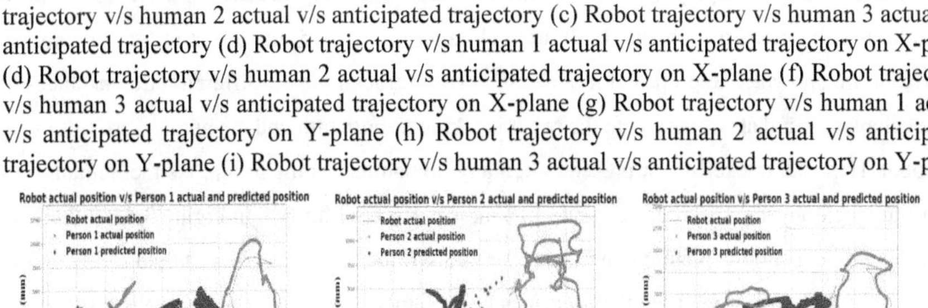

(a) (b) (c)

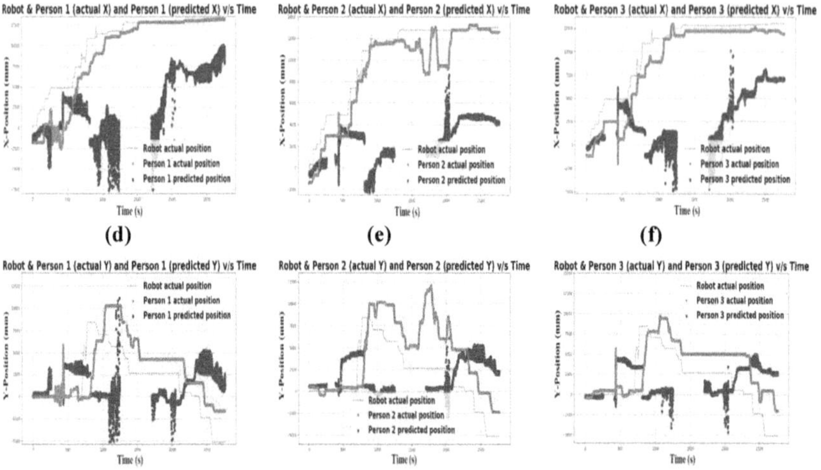

(g) (h) (i)

Figure 3.43: Scenario ID: III 3D trajectory anticipation on a different plane using a LSTM (without re-initialisation) (a) Robot trajectory v/s human 1 actual v/s anticipated trajectory (b) Robot trajectory v/s human 2 actual v/s anticipated trajectory (c) Robot trajectory v/s human 3 actual v/s anticipated trajectory (d) Robot trajectory v/s human 1 actual v/s anticipated trajectory on X-plane (d) Robot trajectory v/s human 2 actual v/s anticipated trajectory on X-plane (f) Robot trajectory v/s human 3 actual v/s anticipated trajectory on X-plane (g) Robot trajectory v/s human 1 actual v/s anticipated trajectory on Y-plane (h) Robot trajectory v/s human 2 actual v/s anticipated trajectory on Y-plane (i) Robot trajectory v/s human 3 actual v/s anticipated trajectory on Y-plane

3.8.3.3 Real-time impression of 3D-PSF model

This thesis has employed the proposed 3D particle social force model for social 3D tracking on multiple behaviours of humans. In Figure 3.44 and Figure 3.45, the different types of human behaviour are shown. The proposed model can perform tracking over different circumstances of these behaviours and at corner where the robot is taking a sharp turn and the humans are following the robot. Here the tracker is not present when any human leaves the group. As the human joins the group, the tracker is re-initialised and tracking is performed further. In Figure 3.44 and Figure 3.45, this phenomenon is pictured and demonstrated. Moreover, the journey started in the indoor environment and after some time the robot reaches a semi-indoor scenario where the presence of sunlight becomes a reason for a change in lightning conditions and the proposed model performs well for tracking, although detection is not error-free here.

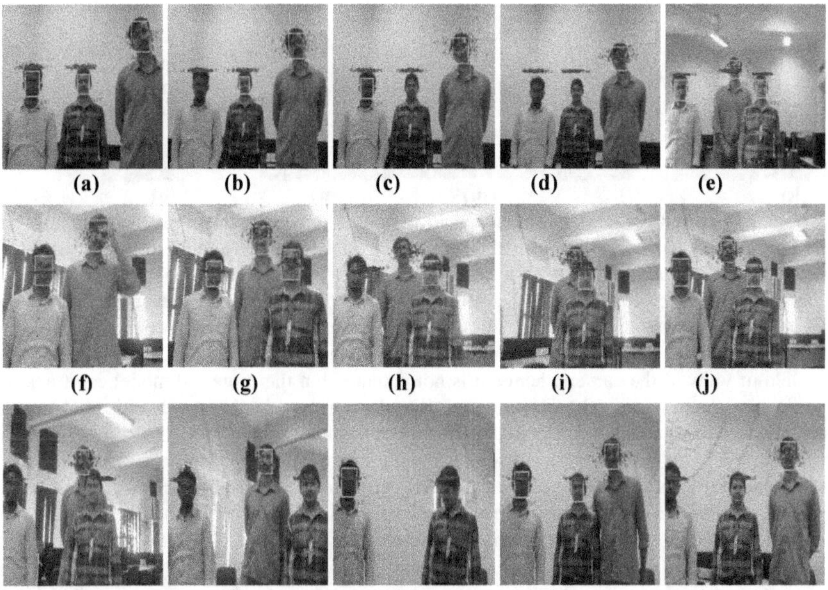

(a) (b) (c) (d) (e)
(f) (g) (h) (i) (j)
(k) (l) (m) (n) (o)

Figure 3.44: Behavioural 3D tracking (a) Every person is detected and tracked by the robot. (b-d) Detection is not possible for every person, but tracking happens. (e) Two humans swapped their position and are following the robot. (f) One person leaves the group, and the tracker is deleted. (g) Person re-joins its group, and the tracker is re-initialised, and tracking starts. (h) A robot is taking a rotation to achieve its goal and here detection is not possible, but the tracking model performs its job. (i-j) One person leaves the group and re-joins. (k-l) A robot continuously takes a sharp turn and humans are also tracked by the 3D tracker model. (m-n) The third person leaves the group and re-joins the group again. (o) Robot is at a corner and determines the human position in 3D.

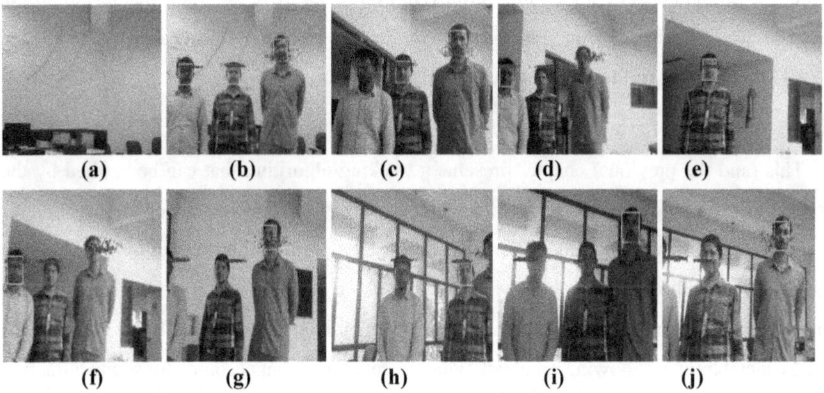

(a) (b) (c) (d) (e)
(f) (g) (h) (i) (j)

(k) (l) (m) (n) (o)

Figure 3.45: Multi-behavioural 3D social tracking (a) All humans of the group have left and the trackers are deleted. (b) Everyone joins the group and trackers are available to track them. (c) Humans' and the robot's position at a corner. (d) Each human is following a robot while maintaining a queue. (e) Two humans have left the group, so the trackers are deleted. (f) Both humans join the group at the corner and follow the robot. Moreover, tracking is being performed. (g) Since the robot is taking a sharp turn and the face of one human is almost cropped due to the limited field of view of the camera, hence it is not detected but the proposed model can track in these critical circumstances. (h) Background illumination is changed and still tracking is performing while one human is cropped. (i-j) The robot is taking a sharp turn to complete its path while tracking. (k) One human is facing forward and the remaining two are backward to the one human. (l) One human is still backward to other ones. (m) Humans adjust themselves due to limited space in the corridor. (n-o) One human has their back to the other due to limited space on the balcony and the robot is taking a sharp turn at the corner.

One of the uniqueness of the work was re-initialisation once a person re-joins a group. In Figure 3.44(f), one person left the group, so the track was deleted. As he joins the group again as shown in Figure 3.44(g), the track is re-initialised, and tracking starts continuously. Another person again left the group and re-joined as shown in Figure 3.44(i-j). In Figure 3.44(k) detection did not happen for one person but tracking was continuously performed. Two persons swapped their position in Figure 3.44(d-e). Further in Figure 3.44(m-n), track for one person was again deleted and re-initialised. Apart from this, all the humans left the group, so the track was deleted as shown in Figure 3.45(a), and as they joined their group (in Figure 3.45(b)), the track was re-initialised, and tracking was performed. Some more track deletion and re-initialisation can be visualised in Figure 3.45. Here we have found 9 number of erratic behaviours.

3.9 Conclusion

This (and the previous) chapter presents a tracking algorithm that can be applied by the social robots to determine the positions of the humans for making informed social decisions for applications such as a robotic guide. The thesis proposed a particle filter model to track humans who are following the mobile robot. Most studies use expensive cameras and stereo vision techniques to track humans in the 3D environment, but this chapter uses only a single budget camera. A social human following robot behaviour is developed that accounts for social attraction and repulsion forces between the people and those with the robot. Here social tracking is performed over humans' behaviour in a group. Humans may show erratic behaviours such as swapping, leaving the group, attending a telephonic call, staying at the previous position due to interest, overtaking, etc. The proposed model can still predict the motion of the human even with a lack of visibility and can also detect humans leaving the group. When any human leaves the group the track is automatically deleted and if the human re-joins, the track is also re-initialised, and tracking continues.

The motivation of the approach came from prior experiments of the author when using mobile robots for social applications. A robot was asked to take a group of visitors on a tour ensuring that all the visitors are visible. However, on several occasions the robot was stuck at a corner and did not move further because the visitors were not visible to the robot. After taking a sharp turn the robot had a wall at the back. There was no space behind the robot so that the visitors could place themselves behind the robot. Some visitors tried hard to get their face visible to make the robot move, which was not possible for all the visitors simultaneously. It was observed that these situations repeatedly occur when the robot reaches at a corner and tries to take a turn for achieving its next goal. Based on these experiments, it was understood that for programming robots for social applications, the robot must always know the person's position and whether the person is in the group. Furthermore, the issues of the robot being stuck around a corner had to be handled in a mechanism independent of the map, which is carried out using the angle to sub-goal. Based on the observations of these initial experiments, the approach assumes that the visitors generally ideally follow the robot and do not leave the group, although they may not always be detected by the robot. The visitors may however sometimes leave the group which must be detected by the robot.

The chapter solves the tracking problem in three phases. In the first phase, an algorithm is used to track vehicles and faces with the subjects at different distances from the robot and recorded with different heights of the camera of the robot. The algorithm performed vehicle tracking in traffic environment using a background subtraction algorithm to detect the vehicles. After background subtraction, the position of every vehicle is estimated by using a Kalman Filter. To detect and track multiple faces a point feature of face is extracted and compared in the remaining frames. The second phase is based on a social particle filter which is used to track the humans for evaluating the "Human following Robot" behaviour. Here the human is available behind the robot and the robot knows the initial position of the human. The subsequent positions are computed by the particle filter model. The initial position of the human is assigned to all the particles. As the robot moves, the humans also move to exactly follow the robot and hence the particles are made to move under the same assumption. The observation model of the particle filter works on the face detection approach. The error between the observed face and the expected face location in the image is used to set the weight of the particle. Resampling is applied after computing the updated weights. The position of the human is estimated from the particle distribution. The third phase extends the second phase to multiple visitors with an ability to detect when any visitor leaves the group. Here a timer is used to measure the duration for which a visitor is not detected. If any visitor of a group is not available for a threshold time duration, it means that the visitor has left the group and the algorithm deletes the tracker which is associated with that visitor. Further, the tracker is re-initialised when the visitor again re-joins the group. While the initial position of the humans is known by the robot, but after leaving the group when the visitors re-join the group, their position is computed by an inverse camera calibration approach using the coordinates of the detected face in the 2D image plane. The approach is benefited by a social potential field algorithm that can accurately predict the position of the humans in the group.

The proposed method gives a small error of 425.26 mm whereas the other baseline approaches give a larger error which is 586.81 mm (RPF), 8105.99 mm (TI), 1045.22 mm (KF), 5378 mm (LSTM with re-initialisation), and 6206.64 mm (LSTM w/o re-initialisation). From an application perspective, if the errors are over 1000 mm, the robot may assume that the visitor is missing or lagging; and may either wait or go towards the human as a social gesture. An error of 425.26 mm on the contrary is acceptable since it does not make the robot take an unwanted decision. Further, if the robot is saying something to the visitors making an eye contact, for errors over 1000 mm the person would be somewhere, and the robot would be looking somewhere else while addressing the person that will be very weird. However, a small error of 425.26 mm by the proposed approach ensures that the robot is nearly always looking at the right direction if an eye contact is to be made. Since the visitors are moving in a group, therefore they will also interact with each other, and moreover, they can leave the group, can attend a telephonic call, overtake another visitor, follow other visitor etc. These social behaviours cannot be handled by several baseline approaches. But the proposed model can handle these behaviours and reduce the error by

27.54% (RPF), 94.76% (TI), 59.32% (KF), 92.09% (LSTM with re-initialisation), and 93.15 % (LSTM w/o re-initialisation).

Further the approach is tested in two different cases, when the robot does not delete the track when circumventing a corner and when the robot may delete the track when circumventing a corner. In this testing it is found that false positives of detecting people leaving the group are rapidly decreased to nearly 0% in all scenarios except for one where there are 77 false positives (that is still better than not handling the corner case by 96.45%). On an average the accuracy of detecting missing visitors is increased by 7% with this approach. Here the algorithm is tested on different scenarios where there is a large diversity in the dataset. The diversity is added by varying the number of corners, distance between two consecutive corners, changes in the workspace, implementation of different erratic social behaviours by the visitors, and changes in the velocity of the visitors. Therefore, in some other cases there are a few false positives. There was one major source of false positive. A goal had a very limited space where every human cannot place themselves behind the robot in such a way the robot can detect them. The algorithm detected that the robot was circumventing a corner and avoided deleting the track for some time. However, the robot soon had to take another turn, increasing the gap between the visitors and the robot itself. After making the turn, it took a lot longer for the humans to place themselves ideally behind the robot that resulted in a false positive.

The approach successfully tracks visitors in a group following the robot. However, there are a few limitations. When the number of visitors increases, the people will need to stand behind each other and the faces of the visitors will be overlapping, and occlusion will happen. So, in this situation detection will not be possible for each visitor. Human guides can still detect people missing a group or lagging, a capability that needs to be incorporated in the proposed algorithm. If the workspace's surface has some small slope, the proposed model will not work because the pose of the camera will change that currently cannot be computed on the fly. The approach uses a separate onboard computing system for the robot and another computing system for the detection and tracking algorithm. Issues regarding connection errors or delays between the two systems are currently not handled. When illumination changes due to the presence of sunlight then detection is not possible. Currently this happens only for a limited time that the tracking algorithm can reject. Persistent detection errors will delete the track assuming that the visitor is missing. Robust detection algorithms therefore need to be developed considering several application scenarios involve outdoor sites with adverse lightning conditions.

Chapter 4

Social Robot Guide

4.1 Introduction

In the recent years, the robotics field has gained more attention and the different wings of social robotics has also created the awareness of many researchers who have presented various applications like robot companion (Dautenhahn et al., 2005), people evacuation (Zelinsky, 1992) and cooperative exploration (Franchi et al., 2007). There are challenging and interesting issues which are involved in the domain of cooperative and social robotics including uncertainty in the manner by which the humans react to a robot, inability to predict the human behaviour, inability to simulate human behaviour for rapid testing, etc. This chapter develops an architectural module for guiding visitors in a group with the robot. Generally, if visitors are moving in a museum or a shopping mall, they do not have any idea of the correct tour path. It may not be possible to employ human guides for monetary reasons. The robots know the correct path of the place and can guide all the visitors in a group. In this application, the robot detects the visitors by their faces. If all the visitors of a group are physically present and detected by the robot, then the robot starts its tour and navigates all the visitors towards different visit sites. As the robot reaches the visit sites, it stops and explains the interesting facts of that place and then moves towards its next visiting site.

This chapter uses a Behavioural Finite State Social Machine modelling to make the robot react to the uncertainties of the human and to socially take the entire group with it, stopping and announcing if absolutely necessary. The robot controller or behaviour may change based on some transition conditions. Unlike a conventional Finite State Machine, every state represents a controller or continuous dynamics within the state, invoked by the controller. Social force and artificial potential field have been used for the physical motion of the robot. The individual behaviours modelled are inspired from the decision making of human tour leaders and the robot judiciously changes its behaviours as per the context. Furthermore, the robot can anticipate the person moving outside field of view and make decisions accordingly. This is the first such attempt to model this critical leading behaviour by the robot under limited field of view settings to the best of the knowledge of the author.

The model of human behaviour has been studied by measurements of position, velocity, flow, etc. and forms a basis to design a social robot behaviour. The human behaviour is chaotic and not easy to predict, especially in complex situations, which makes it hard to model the behaviour. It is assumed that the humans generally move in the same direction; and they fluctuate or suddenly change direction towards goal at narrow crossing areas. These circumstances happen due to non-linear interaction among the human in a group. The problem of finding the potential force to achieve the given behaviour is a very challenging and open issue and mostly non-decidable. A hierarchical methodology (Reif et al., 1999) has evolved to map and model the potential force to achieve the given behaviour of the human.

This thesis has developed a framework for navigation of visitors in a group as briefly shown in Figure 4.1. Visitors are detected by their faces and as all visitors are detected by the robot, the robot continues the tour. If any visitor is missing, then the robot de-accelerates and decreases its velocity from current velocity to zero. This marks that the robot is not sure whether the missing visitor cannot be seen due to uncertainty, cannot be seen as it is out of the field of view, or the visitor is just temporarily missing intentionally. Slowing down is similar to the human gesture of letting the visitor catch up. Consequently, if the missing visitor is again detected by the robot while slowing down, then again the robot accelerates and continues the navigation. However, if the missing visitor is not found for a long period of time, the robot is sure that the visitor is lost or has left the group. In such a case, the robot will make an announcement for the missing visitors, after the visitors are missing for sufficient time. Therefore, if the visitor is missing for a certain duration of time, then the robot stops and waits for the visitor. If the visitor reappears and is detected by the robot, then the robot sends an acknowledgement thorough an announcement and continues the tour.

In the path of the robot, there will be numerous sharp turns because workspaces are designed for humans who generally have no problems in making swift turns at walking speeds. Such turns cause a problem for the robot. When the robot is encountering a sharp turn, it will not be able to see the visitors who will be outside the field of view. A naïve controller may interpret that the visitors are missing and may make the robot stop. This complicates the problem as the robot may stop at a sharp bend which is not wide enough to accommodate visitors to the rear. However, the robot must realise that the visitors are outside the field of view instead. Similarly, even after the robot has successfully completed the sharp turn, the visitors following will take time to account for the same turn and it may undesirable for the robot to stop.

Conventional social robot applications assume that perfect information of the humans is available through sensors. This is not always possible due to sensor noise, limited field of view and sudden changes in the robot's/visitor's orientation. The major novelty in this approach is that the two components are combined, wherein the robot socially responds to the situation so as to mitigate uncertainties in cooperation with the humans. This happens by first slowing down, giving a soft signal to the human; thereafter stopping and subsequently making an announcement, letting the human know the problem. The human then cooperates and comes within the view to mitigate uncertainties.

The second novelty of this approach is using the contextual information from the navigation modules to reason for uncertainties at the tracking and mapping levels. When the robot makes a sharp turn, it is known that the uncertainties will be large due to the visitors being out of field of view. Similarly, when the visitors are potentially navigation a corner, the uncertainties are bound to be large. The robot in such cases anticipates uncertainties and adapts its social response by moving rather than stopping, which is contradictory to the first behaviour.

The third novelty of the approach is at the architectural level. A robotics problem is typically solved in a multi-layer hierarchy. A challenge here is that each component also accounts the social interaction and feedback. The problem is solved as three concurrent behaviours, the reactive navigation mechanism that uses social potential field; the uncertainty-based speed management module that restricts the robot's speed based on specific context and uncertainties; and the higher-level sequencer that operates the overall robotic tour. Both speed manager and sequencer have audio feedback to the visitors in extreme cases and coordinate so as not to pass simultaneous audio feedback.

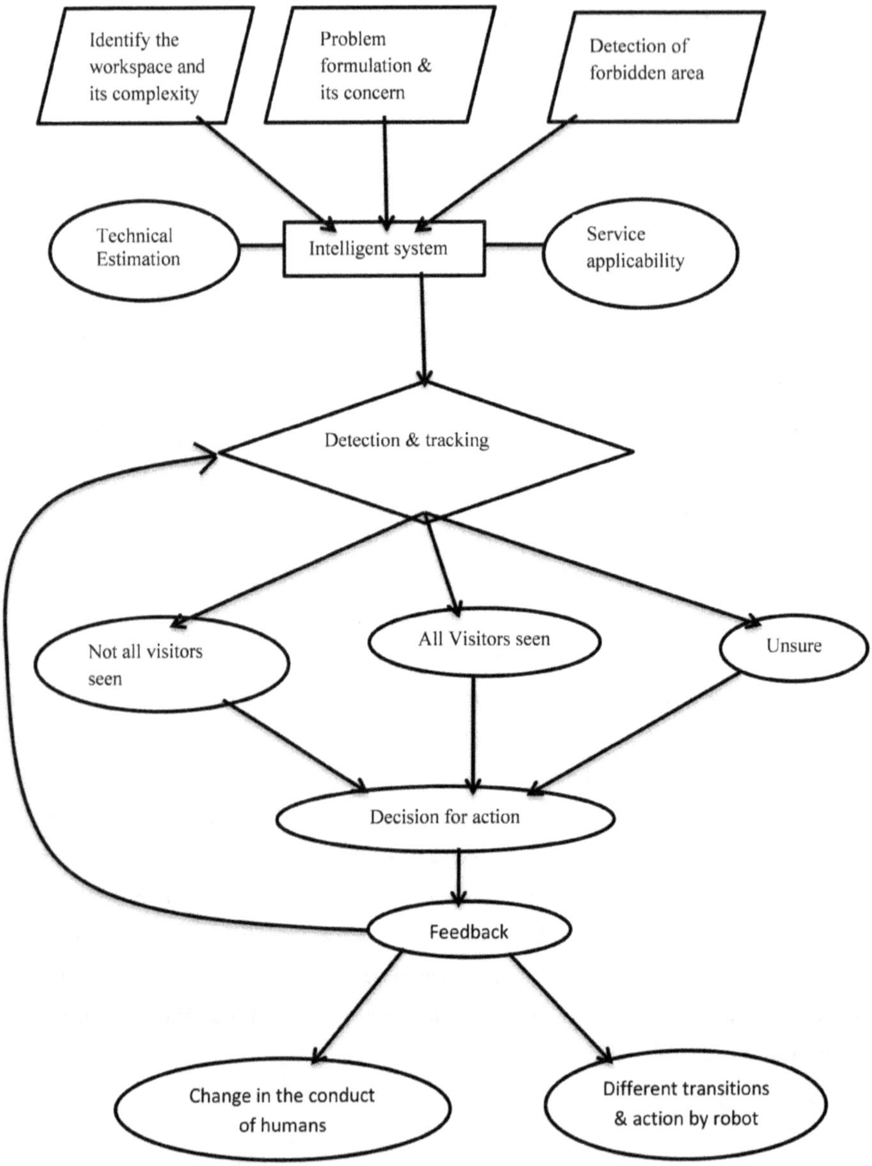

Figure 4.1: Architecture of behaviour change and robot action plan

The key performance indicators are as follows:

- **i.** Ability (and time taken) to detect people leaving the group and stopping, considering that the tour will only continue when all the people are available in the group.
- **ii.** Ability (and time taken) to detect people re-joining the group and resuming the tour.
- **iii.** Comfort factor defined as the maintenance of social distance with the people of the group.
- **iv.** Total time to complete the tour including the time when someone is missing from the group
- **v.** Total time to complete the tour excluding the time when someone is missing from the group

The approach was tested on the Pioneer LX robot at the Centre of Intelligent Robotics of the Indian Institute of Information Technology, Allahabad. The robot was made to take small groups on a tour, which it did well. The approach was also tested with a baseline approach. The baseline approach did not make the robot socially react by slowing down and use the contextual information. The proposed approach was socially better and resulted in tours completing earlier due to absence of false positives.

In this chapter, section 4.2 summarises the related works carried out in the previous years in the literature. Section 4.3 explains the initial model of a social robot guide named as Behavioural Stop and Wait (BSW) machine. Section 4.4 provides a design philosophy which is applicable for social robot guide such as slowing down behaviour of a robot, handling sharp turns and corners etc. The proposed Behavioural Finite State Social Machine (BFSSM) and its algorithm is explained in section 4.5. Section 4.6 describes the implementation details like visit sites, map, system requirements and hardware used. Experiment evaluation and result analysis of each behaviour based social robot guide machines are discussed in the section 4.7. Section 4.8 gives the conclusion of this chapter.

4.2 Related Work

Some recent research studies used a robot as a co-worker exhibiting social human interaction (Ishiguro et al., 2001; Hornby et al., 2000). Different kinds of investigations have been carried out as to the human's attitude towards the robots and their awareness of robots, as an example robotic dog AIBO (Kaplan, 2000), an autonomous robot built for entertainment and attraction purpose. It has an ability to analyse emotions, learning and developing its skills, can show its feelings, and can communicate and interact with humans and children.

Pransky (2004) proposed different applications of a robot for the future by explaining the pros and cons of this type of working robot. The 'Robotic Butler/Maid' accomplishes domestic task, but also causes problems at home making people feel inessential. A "Robot Nanny" (Pransky, 2004) was also introduced that could play with the kids; however, also inhibited them from meaningful communication with human and made communication natural. Vaughan et al. (2000) designed a complete robot system with the capability to gather together a flock of ducks and carry out the task to safely deliver them to a predefined goal.

Various studies have been performed on the robots to guide people in hospitals (Feil-Seifer and Mataric, 2005), railway-stations (Hayashi et al., 2007), museums (Burgard et al., 1998), and shopping malls (Shiomi et al., 2009). Sisbot et al. (2005) examined and analysed robots planning routes as per human preference. Suppose there are multiple routes from one place to another and people opt for any specific route according to their preference, then the robots were also able to understand the behaviour and account for user preference in their route.

Shiomi et al. (2010) developed a group attention control system for the robots to interact with the people in a group (Shiomi et al., 2008). If the robot moves in a backward or forward direction as the people move around it, then the human robot interaction also changes (Shiomi et al., 2010).

Martinez-Garcia et al. (2005) performed a qualitative analysis of the motion of people and multiple robots. However, this system is not applicable in a realistic situation such as the presence of an obstacle, or straying away from their group. These challenges and the motion of the autonomous robot are handled by various heuristics that create a realistic system. Garrell et al. (2009) built the robotic workplace, in which the robot had to operate a guidance task in an open and unbounded area in the presence of obstacles. The algorithm could also find the orientation, velocity, and position of the people and robots, as well as the position of the obstacle in that workplace. It could also predict the intersection of the human with the obstacle and detect the person who is leaving or straying from the group using a Particle filter (Arulampalam et al., 2002).

Detection and sensing of humans who are moving (Rodriguez et al., 2009; Luber et al., 2010; Kratz and Nishino, 2012) has been examined in the area of computer vision and robotics and these approaches depend upon a priori motion model which is typically based on globally optimised parameters for every trajectory during the plenary sequel data. A probabilistic velocity obstacle approach (Fulgenzi et al. 2007) has been combined with the dynamic occupancy grid for the robot navigation with the assumption that the obstacles have a constant linear velocity. Toit and Burdick (2010) designed a robot planning framework which provided the human's anticipated future location and this information also reduced the belief state uncertainty. Other works (Svenstrup et al., 2010; Pradhan et al. 2011) used potential based methodology. For instance, the navigation performance was improved and high-fidelity human motion model was developed that controlled predictive uncertainty (Helble and Cameron, 2007; Thompson et al., 2009; Large et al. 2004; Bennewitz et al., 2005).

In the robot guidance system, the potential function methodology provides an elegant paradigm to specify multiple constrains and goals (Khatib, 1985). There are a lot of applications of potential field in motion planning, navigation and multi robot coordination. Global behaviour of the robot and human can be examined through the superposition of various potential functions (Khatib, 1985) such as motion to goal behaviour, obstacle avoidance etc.

The use of artificial behaviour was demonstrated on a simulation of a school of fish and flocks of birds in the field of computer graphics (Reynolds, 1987). A simple egocentric model was developed for the flocking which was instantiated in every associated member of the simulated set of birds. Reynolds (1987) demonstrated formation of complete and close group behaviour, although individual agents could only sense their local surrounding and very close neighbours. Multi robot behaviours have also achieved recent awareness due to their stability and dynamics (Chen and Luh, 1994; Wang, 1994). The work and experiments were based on the analysis of group dynamics and their stability; however, obstacle avoidance cannot be achieved using this approach.

There exists related works that use the signals in the workspace to enhance and support the navigation scheme in robotics. Here some methods use reference points or specific marks to initially localise the robot in the workspace; although in some other methods an artificial vision approach has been applied to visualise them. Mata et al. (2001) used a visual perception and recogniser system for the indoor workspace which had the utility to check the identity at the office. To enhance the utility of the mobile robot, a topological navigation strategy (Wen et al., 2009) was developed. This topological navigation approach was also applied to the indoor workspace where an artificial mark was used and this artificial mark was called as the Mobile Robot (MR) code. In addition, a topological map was also required in this approach. It stored much less information. This was major a drawback of this method and also its dependency on the lighting situation.

A large amount of research is being undertaken into robot navigation which elaborates that robots are capable of navigating in challenging and difficult environments (Burgard et al., 1999).

Few researchers have started to study how a robot might adjust its velocity when moving next to a human, in environments such as library (Sviestins et al., 2007). Fidelity and security are the major factors for the successful establishment of the mobile robots in a human abode area. In various literature, the security and safety of the robot is confirmed by preventing the people from approaching the mobile robots.

The approaches in the literature largely assume a reasonable knowledge of the human position and delve into the design of social behaviours as a response to the human. These are also adapted to specific applications. However, the robots when operating in such environments are subjected to uncertainties due to their limited field of view, vision uncertainties, need for a sudden turn, unanticipated motion of the human, etc. While the literature develops better vision models to mitigate such uncertainties, it is also important to develop a socially "aware" to adapt the robot behaviour to act in uncertain conditions and to convey the same to the humans. Such adaptations of behaviour have not been attempted in the literature. Since the response is specific to the context and application, this chapter takes the specific application of a robot operating as a tour guide.

4.3 Behavioural Stop & Wait Machine: An initial model for the robot guide

A requirement from a robotic guide is that the robot maintains a social distance from the visitors. If this is not the case, the robot must wait as the visitors may be lagging behind. The waiting behaviour is the same irrespective of whether the visitors are not visible or when the robot is not maintaining a distance. Let d_{soc} be the distance that must be maintained between the robot and the visitors. A realisation from the social experiments is that this distance varies by a very large factor between different visitors, between different moods of the visitors, and between different visit sites. Further, not all visitors may be at comfortable distances and will have mutual social forces between each other, which affects the placement and hence the distance. Since the social distance d_{soc} can only be approximated within a range $d_{soc} \sim [d^{min}_{soc}\, d^{max}_{soc}]$, it is suggestive to have an estimate of the distance between the robot and the visitor.

The closer the face is to the camera (and the robot), the larger is the area of the face and vice versa. So, the area of the face is inversely proportional to the distance of the visitor from the robot. Further assuming the aspect ratio of the human face to be nearly constant, the length and the width of the face is inversely proportional to the distance. This means that the limits of the social distance $[d^{min}_{soc}\, d^{max}_{soc}]$ can be translated into limits on the length of the face $[\eta^{min}_{x},\eta^{max}_{x}]$ in the image and the width of the human face $[\eta^{min}_{y},\eta^{max}_{y}]$ in the image.

4.3.1 Algorithmic Design

The best mechanism to understand the navigation principle is to model the behaviour as a Behavioural Stop & Wait (BSW) Machine. The machine enables the modelling of different behaviours that the robot displays at different phases of time, and to further model the conditions over which the robot behaviour shows a transition. The complete model is shown in Figure 4.2.

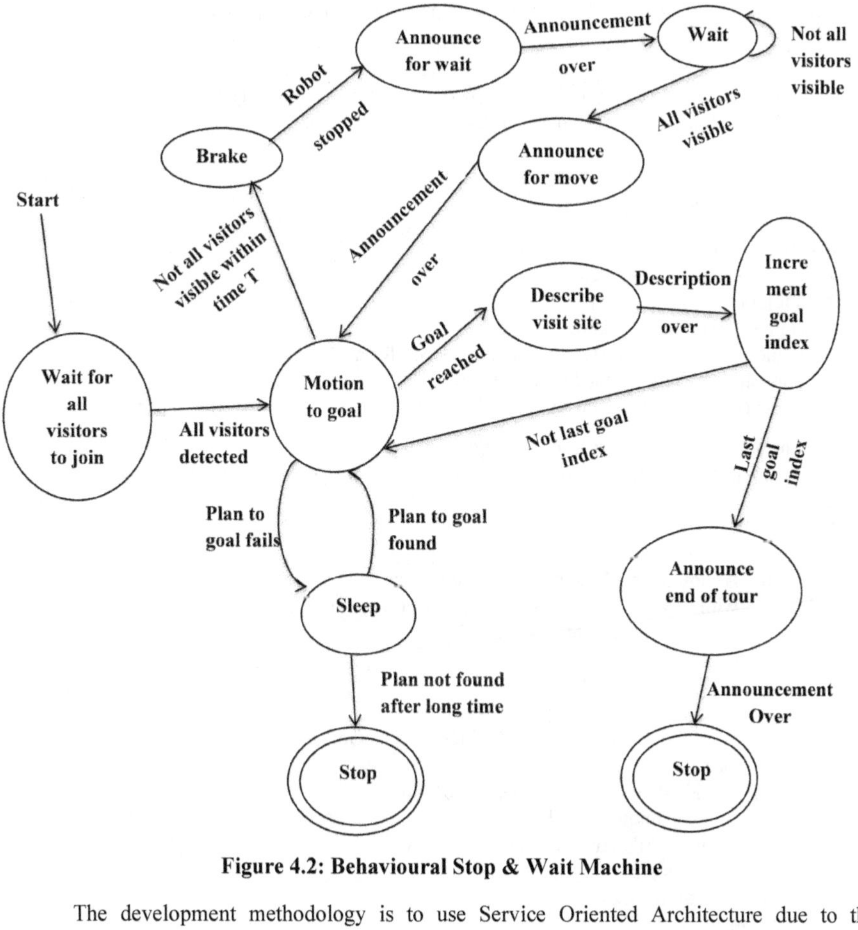

Figure 4.2: Behavioural Stop & Wait Machine

The development methodology is to use Service Oriented Architecture due to the modularity of the different components and naturally suiting the nature of the Robot Operating System, the framework used for the development. The application of Service Oriented

Architecture does not easily translate to the BSW Machine. Further, many linear components of the BSW Machine are combined into procedural units while developing.

Algorithm 4.1: Face detection Algorithm

1. Initialise P-Net, R-net, and O-Net
2. Import and initialise the camera
3. While(tour is operating)
    a. I ← Image from camera
    b. F ← Detect the face by the convolution neural network which is P-Net, R-Net, and O-Net on I with minimum size $[\eta^{min}_{x}, \eta^{min}_{y}]$
    c. F ← Select faces strictly within the image
    d. if (length(F) = noVisitors ∧ (width(f) ≥ η^{max}_{x} ∧ height(f) ≥ η^{max}_{y}) ∀ f ∈ F
        i. visitorsChannel.publish(true)
    e. else
        i. visitorsChannel.publish(false)
    f. Display image I and faces F

Algorithm 4.2: Robot motion algorithm

1. sites ← sequence of sites to visit
2. siteIndex ← 1
3. state ← 'wait for all visitors to join'
4. while 'all visitors visible' not received, sleep for a brief time
5. publish site[siteIndex].location as Goal
6. while (state ≠ stop), sleep for a brief period

Algorithm 4.3: Goal Reached Behaviour(result)
(callback called immediately when a goal is reached or plan fails, result stores the planner's output)

1. if result = "success"
    a. state ← 'describe visit site'
    b. Announce site[siteIndex].announcement
    c. while announcement is not over, sleep for a brief period
    d. state ← 'Increment goal index'
    e. siteIndex ← siteIndex + 1
    f. if siteIndex≤length(site)
        i. state ← 'Motion to goal'
        ii. publish site[siteIndex].location as Goal
    g. else
        i. state ← 'Announce end of tour'
        ii. announce end of tour message
        iii. while announcement is not over, sleep for a brief period
        iv. state ← 'Stop'
    h. Reset failTimer
2. else if result = "fail"
    a. state ← 'sleep'
    b. update failTimer
    c. if failTimer≥T_{fail}, state← 'Stop'

Many components use the face detection results as sensory information. Algorithm 4.1 is the face detection algorithm. The CNN network model for face detection was detailed in Chapter 2. Algorithm 4.2 implements the initial motion of the robot. The motion thereafter is by a set of registered call backs only. Algorithm 4.3 implements the behaviour when the goal is reached. The motion planning library, not a part of the algorithm, calls the callback function whenever the goal is reached successfully or the plan fails. The "result" variable stores the results received from the planner. The other module is to stop and wait until all visitors are visible. The face detection algorithm keeps publishing the visitor status and therefore the corresponding callback that receives the visitor status is always active, where the logic is implemented. This can also be visualised as a continuous loop that maintains the state and invokes the corresponding behaviour. The callback implementation is given by Algorithm 4.4.

Algorithm 4.4: Waiting Behaviour(faceDetection)
(callback called immediately and every time when face detection messages are received)

1. if faceDetection = 'not all visitors not visible' ∧ state = 'Motion to goal' ∧ notFaceVisibleTimer > T
    a. state ← 'Brake'
    b. publish 'cancel motion'
    c. while(velocity>0) sleep for a brief period
    d. state ← 'Announce for wait'
    e. announce wait message
    f. while announcement is not over, sleep for a brief period
    g. state ← 'Wait'
2. if faceDetection = 'not all visitors not visible' ∧ state = 'Wait'
    a. state ← 'Announce for move'
    b. announce move message
    c. while announcement is not over, sleep for a brief period
    d. publish site[siteIndex].location as Goal
    e. state ← 'Motion to goal'
3. if faceDetection = 'all visitors not visible' Reset notFaceVisibleTimer
4. if faceDetection = 'not all visitors not visible', update notFaceVisibleTimer

4.3.2 Limitations of the initial model for the robot guide

The initial model for the robot guide has some shortcoming which are as follows:

i. The robot moves with a constant velocity and therefore when the visitors are not detected for a sufficient time, then the robot suddenly stops. It is possible that visitors may be available behind the robot but not currently within the field of view.
ii. Most of the time visitors may be available behind the robot but due to contextual reasons such as the robot making a sharp turn, the robot does not detect them and makes a false

positive. The little delay in stopping does not give enough time to rectify the context, while increasing this time threshold is not a solution since the robot will continue the journey for enough time even after the people have actually left the group.

iii. If the person repeatedly wants to go slower than the robot, or repeatedly intends to stay for a larger time than what is given by the robot, then the robot will have to often stop. A more socially acceptable response is to navigate slower to indicate that the person should catchup, while go at a normal speed once the person starts to follow.

iv. The robot journey duration time is very long because of these limitations.

4.4 Design Philosophy

This chapter now designs the best strategy with the following principles: (i) while the robot is navigating all the visitors should be in a group and if any visitor leaves the group, then the robot should not continue the tour and must stop and wait for the visitor to join the group, (ii) the robot may need to make sharp turns with the visitors following. The robot should tackle the sharp turns itself and allow the visitors to turn the same in a manner of their liking. Visibility is likely to be lost during this time. The robot should not wait for the visitors to be exactly at the rear of the robot at every instant of time. (iii) The false positives should be as few as possible, or the robot must never wait if all visitors are approximately around however the lack of visibility is due to contextual reasons or errors in the detection algorithm. (iv) If the robot detects that a visitor is missing, the robot must attempt regrouping by making an announcement. (v) Only one robot is available for navigating the visitors.

In the robotic tour only one robot as a captain and a group of visitors are present. The captain robot finds a path and moves from one visit sites to the next along with the visitors. This robot also watches the visitor's activity behind it. If any visitor leaves the group, then the robot de-accelerates until its velocity becomes zero and thereafter it makes an announcement for the visitors to join a group and waits for them to assemble. After the visitors join the group, the robot moves forward for its next task. Here the robot also reaches corners which require it to make a sharp turn to get to the goal. The robot must make such turns during which time the visitors may not be visible due to the limited field of view of the camera.

4.4.1 Navigation Strategy Modelling

Consider the robot's pose in the real world as q_R, the rear camera at the robot with a pose q_C and a visitor's (specifically the visitor's face) position as q_V. The camera is assumed to be at the centre of the robot, with an additional height, looking at the rear, which gives the transformation between the robot and the camera (T_R^C) as Equation (4.1);

$$T_R^C = \begin{bmatrix} -1 & 0 & 0 & 0 \\ 0 & -1 & 0 & 0 \\ 0 & 0 & 1 & h_C \\ 0 & 0 & 0 & 1 \end{bmatrix} \tag{4.1}$$

Here h_C is the height of the camera. The -1 denotes that the camera is looking backward, or a Z axis rotation of π. The transformation between the visitor and the robot ($T_V^R(q_V, q_R)$) can be obtained with a knowledge of their global positions. The visitor as seen in the image of the camera is given by Equation (4.2);

$$q_I = \zeta T_R^C T_V^R q_v \tag{4.2}$$

Here ζ is the distortion corrected calibration matrix. However, since the camera has a limited field of view, the visitor will be visible in the image only when the visitor is within the field of view, given by the function $f_V(\zeta T_R^C T_V^R q_v)$, which takes the visitor's projection into the camera as an input and checks for bounds.

Ideally, the visitor is supposed to follow the robot, whose generic control law is given by Equation (4.3);

$$\begin{bmatrix} v_V \\ \omega_V \end{bmatrix} = \begin{bmatrix} K_1(d(q_v, q_R) - r) \\ K_2(atan2(y_R - y_V, x_R - x_V) - \theta_V) \end{bmatrix} \tag{4.3}$$

Here K_1 and K_2 are the gain constants. $d(q_V, q_R)$ is the distance between the robot and the visitor. r is the sum of radii of the robot and visitor and a little extra safety distance, θ_V is the orientation of the visitor, (x_R, y_R) is the position of the robot and (x_V, y_V) is the position of the visitor.

Under the ideal circumstances, if the robot is stationary, the visitor approaches the robot and stops at a comforting distance called as the personal space. As the robot moves the visitor follows the robot as long as the robot's speed is smaller than the preferred speed of the visitor. The ideal following behaviour may be disrupted by obstacles. However, the visitor may not adhere to the ideal control laws. The visitor can get attracted by other things temporarily, leave for reasons like a phone call, etc. The robot must adjust to the same. One possible mechanism is to have an attractive force between the robot and the visitor, which may make the robot follow the person on the deviated track, a behaviour which is inadmissible for the application.

The other mechanism is to make the robot wait when the visitor has temporarily left the group and resume the tour when the visitor re-joins. However, the challenge is to correctly identify the intention of the person. A few steps away from the robot could be due to an interesting landmark of interest or the person avoiding an obstacle, when the robot should not wait; while it could also be due to the visitor leaving to buy snacks, when the robot should wait. This chapter considers the problem when the person is out of bounds of the camera and cannot be tracked, or the bounds of the view function $f_V(\zeta T_R^C T_V^R q_v)$ do not hold. The view function depends upon the pose of the visitor and the robot both, discussed separately in the next 2 sub-sections.

4.4.2 Slowing Down of the Robot

This sub-section is devoted to the out of view conditions $f_V(\zeta T_R^C T_V^R q_v)$ due to the visitor alone. In general, the robot navigates the visitors if all the visitors follow the robot. Conventionally, if any visitor does not follow the robot, the robot stops suddenly. Moreover, the person may be present and not within the camera's field of view. If the robot suddenly stops, the person does not know about the reason and the person shall never come within the view. This thesis solves this issue in this navigation strategy so that robot does not produce any jerk. This thesis has designed the slowing behaviour of the robot. The robot navigates with its normal speed and suddenly if any visitor is not detected, then the robot does not stop but reduces its speed until its speed becomes zero, when it stops. Consequently, if the missing visitor is detected again by the robot, then again the robot increases its speed to normal. The slowing down behaviour lets the robot overcome the false positives of detection. Furthermore, since the robot is moving, it is possible that the missing visitor re-appears to the robot and the presence of the visitor is re-ascertained, since the camera and its orientation changes with the robot. The robot acceleration or deceleration depends upon the availability of the visitors behind the robot.

Within the limited and uncertain sensing, the decision to wait and affirm that the visitor has temporarily left the group is when it is not seen for a threshold amount of time (t_{stop}), or Equation (4.4)

$$left_V = true \text{ iff} f_V\left(\zeta T_R^C T_V^R q_v(t)\right) = false \ \forall t - t_{stop} \leq t \leq T \tag{4.4}$$

Here T is the current time. The probability that the visitor has temporarily left the group is given by Equation (4.5), which has a value of 1 when Equation (4.4) holds and slowly increases from 0 to 1.

$$P(left_V) = \min\left(\frac{T - t_{lastSeenV}}{t_{stop}}, 1\right) \quad (4.5)$$

$$t_{lastSeenV} = \arg\min_{v} \arg\max_{f_V\left(\zeta T_R^C T_V^R q_v(t)\right) = true} t$$

The minimisation over v is to make a decision based on the visitor which is not seen for the largest time. Instead of making a sudden decision to stop abruptly when it is ascertained that the visitor is missing, the behaviour is smoothed by allowing the robot to slow down as its confidence that the visitor has left increases. The maximum permissible speed v_R^{pref} is hence given by Equation (4.6)

$$v_R^{pref} = v^{max} P(left_V) = v^{max} \max\left(\frac{T - t_{lastSeenV}}{t_{stop}}, 0\right) \quad (4.6)$$

Here v^{max} is the maximum allowable operating speed of the robot without considering the uncertainty constraints.

4.4.3 Handling sharp robot turns and corners

The other case when the view function $f_V(\zeta T_R^C T_V^R q_v)$ does not hold is due to the motion of the robot. While the robot typically operates at a low speed causing little change in position, the turn of the robot or change in orientation can place the visitor outside the field of view. As the robot moves, it may have to take sharp turns to avoid obstacles. In such cases, the visitors will move out of the field of view. Even after the robot completes the turn the visitors will still not be in the field of view since they would be themselves making the same turn. In such cases, the robot should not slow or stop, which will create a false positive as the visitors are still there. The stopping of the robot will imply that the visitors need to quickly come behind the robot, which may not be possible as there may be no space in the corner. In this navigation strategy, the proposed approach has solved this issue. In such a case, the robot does not have to account for the visitors. Although at every stage except the corner robot is looking behind it to know the visitors' appearance.

The problem is again to detect such a situation and to take corrective actions in the navigation strategy of the robot. It may initially appear that a high angular velocity is a good indicator of sharp turns. However, numerous control mechanisms may produce fluctuating control signals, which can be hard to interpret for such decision making. Hence, the angle to the sub-goal is used as a measure to detect sharp turns and corners.

The path of the robot to the goal or visit sites may be broken into multiple sub-goals. As soon as the robot draws near to a sub-goal, the next sub-goal in the path to goal acts as the current sub-goal. This continues till the last sub-goal (always the goal) is the current sub-goal. This is as per the working of a fusion of deliberative and reactive navigation mechanism.

Let s be the current sub-goal in pursuit of visit site g, which is the ith sub-goal in the deliberative path from source to visit site g. A typical motion of the robot is to travel in a straight line from the current position to the sub-goal, in which case the robot nearly always faces the sub-goal. Any large deviation from the sub-goal means that the robot had to severely disorient itself to avoid an obstacle. Alternatively, it means that the current sub-goal changed and the new sub-goal requires a sudden change in the orientation of the robot, denoting a corner. The detection of a sharp turn or corner is hence given by Equation (4.7).

$$turn = true\ iff\ \cos(\theta_e) < \eta \qquad (4.7)$$

$$\theta_e = atan2(y_s - y_R, x_s - x_R) - \theta_R$$

Here (x_s, y_s) denotes the coordinates of the sub-goal and (x_R, y_R) denotes the coordinates of the robot. η is a constant.

If the robot is found to be turning, no slowing or stopping constraints are applied and the robot continues to move until the turning conditions cease to apply. After the turning conditions cease to apply, the robot is said to have accounted for the turn and the tracking continues. It may be argued that the tracking further needs to be off considering even after the robot accomplishes the turn, the visitors will need time to accomplish the same turn, during which time they will not be visible to the robot. This will naturally be handled by the robot as per the slowing conditions.

4.5 Algorithm Design

4.5.1 Overall Architecture

The overall architecture has three distinct modules which work and share information with each other as shown in Figure 4.3. The first module is the speed manager that manages the maximum speed of the robot and also provides the ability for the robot to accelerate for its acceleration and decelerate. If visitors are continuously detected by the robot, it will move with maximum velocity, while if any person is not detected, the velocity of robot decreases. As the visitors again appear, the velocity increases. The robot will only stop when visitors are not detected for a certain time. This module also handles the turning cases, when the robot does not react to missing visitors on making a sharp turn. This module also sends audio feedback to the visitors, telling them if a visitor is missing or if the missing visitor is now seen and the group must continue.

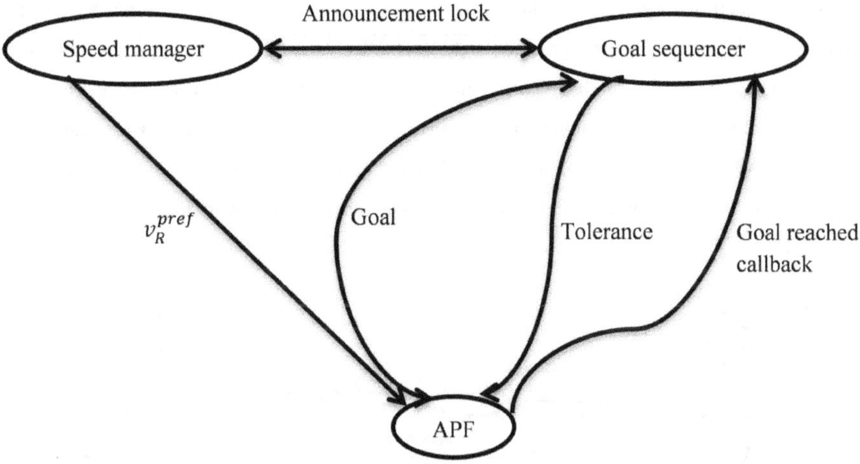

Figure 4.3: Goal sequencer

The second module is the goal sequencer. Here the different visit sites or goals of the environment (workspace map) are defined. The module sequences different visit sites. Reaching each visit site is a classic motion planning problem, which is broken into multiple sub-goals. On reaching each visit site, the robot makes the announcements giving information about the site as a tour guide.

An announcement lock system is also built into this scheme. Both the speed manager module and the sequencer module run in parallel and use the speaker as a resource for giving audio feedback to the visitors for different reasons. It is possible for both audio messages to be invoked together. A module using the speaker must first lock it and unlock it after the announcement. This locking status is shared and communicated between modules, so, when one module uses the speaker for announcement, the other module will not be able to announce anything and will have to wait for the other module to finish the announcement execution.

The Artificial Potential Field (APF) is the third module which acts as a controller for the robot. Here potential function and force are applied on the robot, which is used to calculate the control signal for the robot. The APF gets its goal from the sequencer and permissible speed limits from the speed manager. The module also keeps a track of the robot position and when it reaches the next sub-goal (or goal) within the set tolerance limit, it gives feedback to the sequencer.

4.5.2 Speed Manager

The speed management module sets an allowable navigation speed for the robot and the module is modelled as a Behaviour Finite state Social Machine (B-FSSM) for the navigation of the robot. The term social is deliberatively added to the term to denote that this module interacts with the humans, both in tracking them and using the same information for decision making, as well as to give them feedback using audio signals. The module empowers the robot to make decisions as to its speed based on the different states as shown in following Figure 4.4.

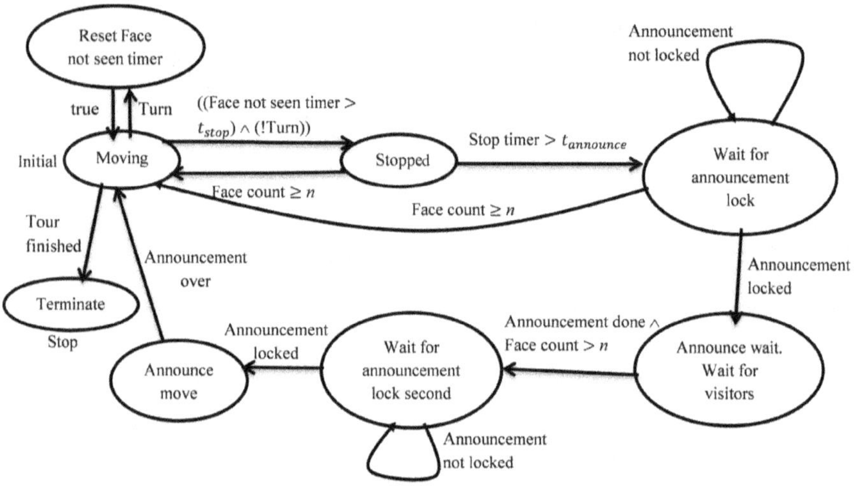

Figure 4.4: Task based B-FSM (Behaviour Finite state Social Machine)

The robot is initially at the moving stage, wherein it moves as per the sequencer and controller. In the BFSSM a timer is assumed to available at every stage. Ideally, the robot keeps moving as per the sequencer and controller. However, if any visitors are not available for a certain time then robot the robot adjusts its speed as per the time when the visitors were not seen. Eventually, after t_{stop} amount of time, if the visitors are still not seen, the robot will stop. This gives a soft visual feedback to the group that something is wrong and if the missing visitor is around, they would come back, in which case the state is again changed to moving. On the contrary, if this soft feedback does not help, after $t_{announce}$ time an audio feedback is provided to attract the attention of the missing visitors. The announcement system first needs to be locked to avoid two modules making an announcement simultaneously, which results in an interim 'waiting for announcement' state. Once all the missing visitors are detected by the robot, the robot makes an announcement and then restarts its journey. The detection of visitors does not happen if the robot is encountering a turn. This repeats till the tour completes.

Algorithm 4.5: Velocity manager

1) Get visitorsStatus from visitorsChannel
2) Get goalStatus from goalChannel
3) Localise Robot
4) state = "Moving"
5) While (goalStatus≠"end")
   i) Get visitorsStatus from visitorsChannel
   ii) Get goalStatus from goalChannel
   iii) T = Current Time
   iv) If (state ="moving")
      (a) If (visitorsStatus $\vee \cos(\theta_e) < \eta$), $t_{lastSeenV} = T$
      (b) $v_R^{pref} = v^{max} \max\left(\frac{T - t_{lastSeenV}}{t_{stop}}, 0\right)$
      (c) If (state="moving" $\wedge\ v_R^{pref} = 0$)
         1. $t_{stopped} = T$
         2. state="stopped"
   v) Else if (state = "Stopped" $\wedge\ v_R^{pref} = 0 \wedge T - t_{stopped} > t_{announce}$)
      (a) State = "Wait for announcement lock"
   vi) Else if (state = "Wait for announcement lock" $\wedge \neg$ announcementLock)
      (a) announcementLock = True
      (b) announcementChannel.publish(announcementLock)
      (c) play(wait message)
      (d) announcementLock = False
      (e) announcementChannel.publish(announcementLock)
      (f) State = "wait for visitors"
   vii) Else if (State = "wait for visitors" $\wedge$ visitorsStatus)
      (a) State = "Wait for announcement lock second"
   viii) Else if (state = "Wait for announcement lock second" $\wedge \neg$ announcementLock)
      (a) announcementLock = True
      (b) announcementChannel.publish(announcementLock)
      (c) play(moving message)
      (d) announcementLock = False
      (e) announcementChannel.publish(announcementLock)
      (f) State = "moving"
   ix) speedChannel.publish(v_R^{pref})

Compressing some of the states of the BFSSM, the simplified algorithm for the velocity manager is given as Algorithm 4.5. The algorithm receives a *visitorStatus* from the face detection

Algorithm 4.1. The status is true if all the visitors are visible and false otherwise. Even though all modules interact using ROS, the passing of information from face detection to the velocity manager is not carried out by ROS due to the latency incurred by ROS.

4.5.3 Goal Sequencer

The sequencer is responsible for ensuring that the robot goes from one visit site to the next and completes the entire tour. At the highest level, the problem involves reaching each visit site, explaining the facts about the visit site, and continuing to go to the next visit site. Let $V=[v_i]$ be the ordered sequence of sites with $v_i.q$ denoting the location of the site v_i and $v_i.t$ denoting the information to be read about site v_i. After all the sites have been visited, the robot reads out the concluding lines for the tour. The problem is hence to sequence all the visit sites and narrate facts about them in a sequence.

Algorithm 4.6: Goal Sequencer

1) Initialise V
2) i = 1 (goal), j=1 (sub-goal)
3) Publish s_{ij} as goal
4) If j = len($v_i.\tau_d$), publish small tolerance (goal), else publish large tolerance (sub-goal)
5) while (state not end of tour), sleep

Algorithm 4.7: Goal Reached Status(status)

(The function is called whenever some data is published on the goal reached status)

1) if result = "success"
   i) j ← j +1
   ii) if(j>len($v_i.\tau_d$))
      (a) while (announcementLock), sleep for a brief period
      (b) announcementLock = True
      (c) announcementChannel.publish(announcementLock)
      (d) Play(v_i.t)
      (e) announcementLock = False
      (f) announcementChannel. publish (announcementLock)
      (g) i←i+1, j←0
         1. if (i>len(V)), goalChannel.publish("end")
         2. else goalChannel.publish(s_{ij})
         3. if(i<len(V) ∧ j = len($v_i.\tau_d$)), toleranceChannel.publish(ε)
         4. else toleranceChannel.publish(E)
   iii) else goalChannel.publish(s_{ij})

The deliberative path is in the form of a sequence of sub-goals or $v_i.\tau_d=[s_{i,j}]$, where $v_i.\tau_d$ denotes the deliberative trajectory to reach the visit site v_i. The robot travels towards the sub-goal. As soon as the robot arrives within the tolerance limits of the sub-goal, the sub-goal is said to have been reached and the next sub-goal (or goal) is loaded. If the robot reaches the goal, the announcement system plays the informational material about the visit site or goal. The overall algorithm is given by Algorithm 4.6 and Algorithm 4.7. Here ε is the accepted tolerance for the goal and E is the accepted tolerance for the sub-goal.

4.5.4 Artificial Potential Field

The robot seeks the sub-goals using the artificial potential field. The robot is equipped with a laser and the laser readings give the information about the obstacles, including the distance from the obstacle and its direction. The repulsive force due to the laser at an angle is φ reporting a distance to obstacle as d_φ, with the robot itself at an angle θ_R is given by Equation (4.8). Here α is the repulsive constant.

$$F_{rep} = -\sum_{\phi=\phi_{min}}^{\phi_{max}} \alpha \frac{1}{d_\phi^2} [\cos(\theta_R + \phi) \sin(\theta_R + \phi)]^T$$

(4.8)

Similarly, the attraction from the sub-goal s (x_s, y_s) with the robot at q_R (x_R, y_R) is given by Equation (4.9);

$$F_{att} = \beta d(q_R, s)[\cos(\theta_s) \sin(\theta_s)]^T$$

(4.9)

$$\theta_s = atan2(y_R - y_s, x_R - x_s)$$

The total force is given by Equation (4.10);

$$F_{total} = F_{att} + F_{rep} \quad (4.10)$$

4.6 Implementation details

This methodology has been implemented on a master ROS framework. Webcam is used to visualise humans and detect their faces. The proposed navigation strategy was tested on a simulator first then on the real robot. The Pioneer LX robot is used to check the proposed methodology. The Centre of Intelligent Robotics at IIIT Allahabad, Prayagraj was the workspace for navigation. Webcam has tightly fixed on a monopod and this monopod was mechanically fixed on the Pioneer LX robot. ROS ARNL and ROS ARIA library provided the basic functionalities of operating the robot.

The first problem was to make a map of the laboratory. In order to do this the robot was first operated in the laboratory. The map is made based on the lidar readings and the wheel encoders of the robot. Thereafter, the different sites of interest were manually marked on the map.

These constitute the visit sites that the robot should visit in the given order. The ordering was carried out manually. The deliberative path between the visit sites was also manually specified.

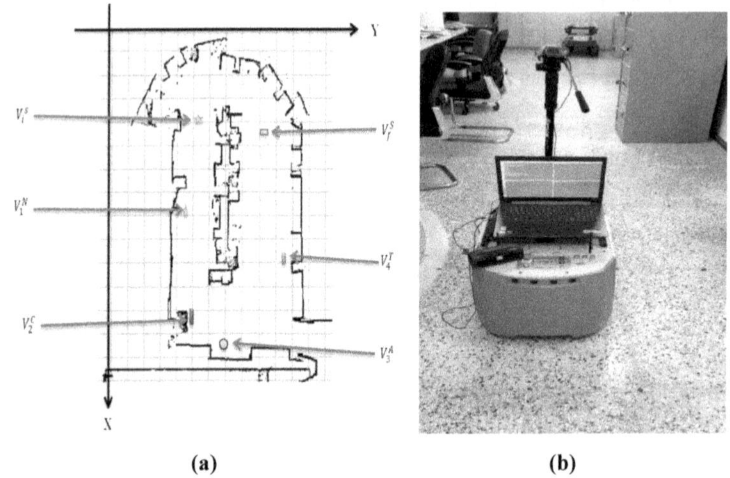

Figure 4.5: (a) Different visititing sites on the workspace (b) Robot setup for visitors navigation

The algorithm was executed in two systems. The first system was a laptop with system configuration 8 GB RAM and core i5 processor. This system did most of the processing including image processing, visitor detection, velocity management and potential field. The other system was the embedded system in the robot which used the commands for the navigation of the robot. Both the systems were connected using the ROS interface.

In the Figure 4.5 (a) the workspace is shown with various visiting sites while Figure 4.5 (b) represents the robot setup through a robot master slave. Descriptions of various visiting sites are as follows:

- V_i^S: This is initial state of the robot or starting point of the robot for visitors' navigation.
- V_1^N: It is first visit site where *NaO* and *AmigoBot* robots are placed.
- V_2^C: An in-house developed autonomous self-driving mini car called Autonomous Robot Testbed
- V_3^A: *Adaptive Modular Active Leg (AMAL)*, an in-house developed prosthetic leg available in the laboratory
- V_4^T: A social robot for interaction with children named as *Tinku*

- V_f^S: Last visiting point or final state of the robot where a social mobile robot called SMART is located.

4.7 Experimental evaluation and result

Real time experiments have been performed on the BSW Machine, and Behavioural Finite State Social Machine. Moroever, addition experiments have been performed for face width and height analysis with different distances.

4.7.1 Social Experiments

In the initial set of social experiments, the visitors were asked to stand at a comfortable distance from the robot. The visitors showed a great disparity in maintaining distances, ranging from fairly small to fairly large. This makes it clear that there is no single social distance and different visitors have different social distances. A part of the problem was also because of the fact that since the visitors had never travelled with a robot before, they did not have any preferences which in the case of humans are developed over a considerable length of time. Therefore, some visitors stood close, while the others were fearful and preferred to keep a distance. The last aim of the social experiment was hence to calculate the range that appreciably covers the acceptable social distances that the visitors keep with the robot and to convert the same into the width and height of the faces as seen in the camera attached to the robot.

To explore the social distances between the robot and visitors, while the robot is guiding in a group, the proposed approach has calculated the height and width of all the faces in different scenarios and multiple distances with multiple visitors. The proposed method determined the height and width of only one visitor when the visitor is at multiple distances from the robot such as 1, 1.5, 1.8, 2, 2.5, 2.7, 3.1, 3.4, 3.5 metres and so on. Similarly, the height and width of all faces with one, two, three, four, five, and six visitors in a group have been determined and tested at multiple distances from the robot. Suppose there are five visitors in a group, and distance between the visitors and the robot are 1.1 metres, 2.1 metres, 1.6 metres, 3.4 metres and 2.8 metres. In this case, the proposed method calculates the height and width of all these five faces the distances from the robot. In each experiment the visitors are asked whether, given the group and the position of other members of the group, they find the distances comfortable or not. This strategy was applied with different number of visitors in a group (like 1, 2, 3, 4, 5, 6) with various distances (for example 0.92, 1, 1.1, 1.4, 1.8, 2, 2.4, 2.7, 3.1 metres etc.) to determine and analyse the height and width of all faces. By doing this it is easy to find a lower and an upper bound of height and width of the faces which would be applicable to the robot to navigate the visitors in a group. The best estimate of the height and width that matches with the comfortable distance as reported by all the visitors is considered.

Table 4.1: Distance Measurements Group 1: Computation of width and height of visitors at different distance from the robot

Scenario ID	Image taken by the camera	Number of visitors (actual)	Number of visitors (predicted)	Visitor ID	Weight of face	Height of face	Distance from robot	Comfort factor
1		2	2	1	15	19	3.18	1
				2	17	21	2.95	
2		2	2	1	17	21	2.95	1
				2	15	18	3.36	
3		2	2	1	16	19	2.95	1
				2	24	30	2.34	
4		2	2	1	16	20	2.95	1
				2	31	37	1.81	
5		2	1	1	17	21	2.88	0
6		1	1	1	11	14	4.21	0
7		2	2	1	19	23	2.95	0
				2	12	14	4.21	
8		2	2	1	17	21	2.88	1
				2	15	19	3.58	
9		2	2	1	27	32	2.10	1
				2	16	21	2.96	
10		2	2	1	37	48	1.37	0
				2	27	33	2.10	
11		1	1	1	14	16	3.97	0

Here some experiments are summarised in Table 4.1 and more experiments are available in Annex A.1 to A.7. The table uses two symbols [0, 1] to state the comfort factor, a key performance indicator for the approach. Comfort factor is used to evaluate the specific distance between the robot and the human. Comfort factor is defined as the maintenance of social distance with the people of the group as judged by the people participating in the experiments. If the value is 0 then the robot is adjudged to have not maintained the required social distance as per the humans, and if it is 1 then the robot is said to have maintained the required social distance.

In Table 4.1 (scenario ID 1), there are two visitors who are present in a group and each visitor has a unique visitor ID as 1 and 2. The robot detects both visitors via their faces and distance between visitor 1 to robot is 3.18 metre and visitor 2 to robot is 2.95 metre. Therefore, the width and height of all the faces will not be the same here since both visitors are at distinct distances from the robot. The width and height of the face of visitor 1 is 15 and 19 while visitor 2 is 17 and 21 respectively. Also, in this scenario the robot's comfort factor is 1, it means that the robot will be moving all the visitors in a group. This complies with the personal observation that they were happy for the robot to move as a guide. The other cases can be similarly understood from the table.

After looking at all the experiments with different visitors, different groups and different distances from the robot and it has found that the 15 minimum and 45 maximum as very good bounds for width and height of the faces, so as to maintain a comfortable distances from the robot. The robot detects the face, and moves towards the visit sites with the group if the width and height of the faces is lies in between the range 15 to 45. This range value (15 to 45) of width and height produces a comfortable distance between visitors and robots as a minimum of 1.2 metres and a maximum 3.5 metres. So, while robot is navigating in a group, the visitor will have to maintain a distance between robot to 1.2 metres to 3.5 metres. Otherwise the robot waits for the visitors. In very less and exceptional cases the value of width and height of the face is larger than the upper bound, when the visitor has moved too close to the robot. The robot will not move, which does not improve the situation, but the probability of this situation arising is very low and the visitors will have to understand this social cue and move backwards. This is unlike the case when the visitor is far behind and the robot waiting gives enough time for the visitor to come and join the group.

4.7.2 Results on a real tour using Behavioural Stop & Wait Machine

The results of the Algorithm 4.1-4.4 are given in Figure 4.6. In Figure 4.6 (a) the robot detected visitors and started its journey and this point was also the initial or starting point of the tour. After a few milliseconds, the robot moved towards its first goal (shown in Figure 4.6 (b)), where the NaO robot and AmigoBot robots were situated. Figure 4.6 (c) shows the robot reaching the first visit site.

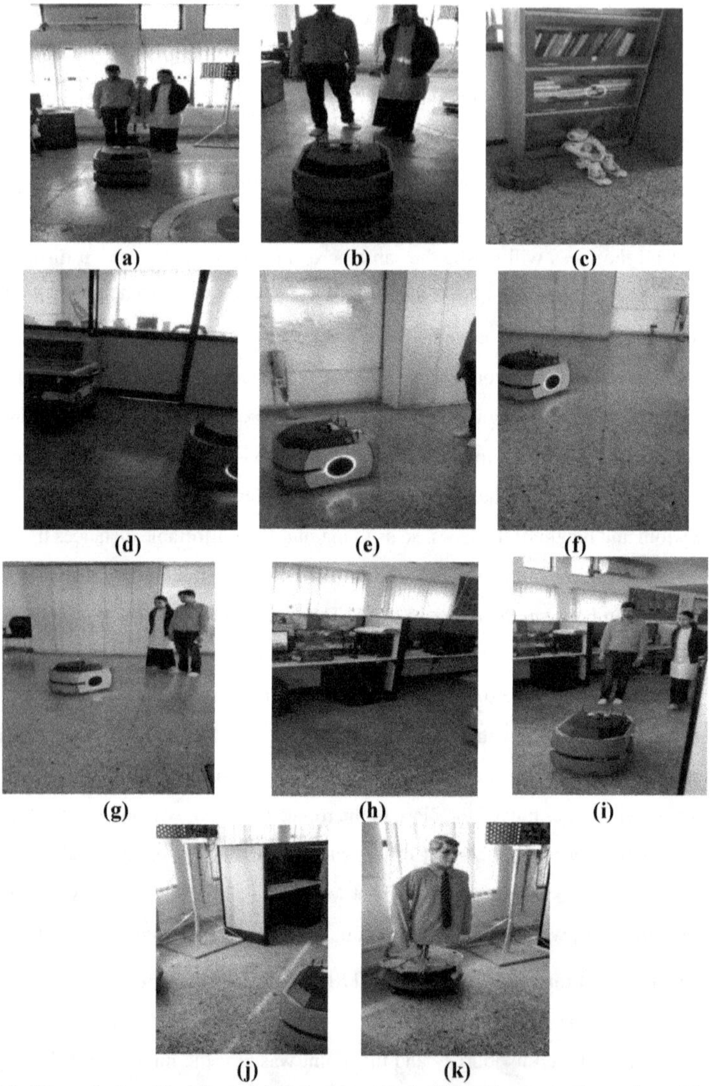

Figure 4.6: The robot guided tour (a) Robot is at its initial state or starting point of the journey and detected the visitors through their faces through the webcam mounted on it. (b) Robot reaches its first visit site, (c) NaO and AmihgoBot are situated at the first visit site of the robot. (d) Robot

has arrived at its second visit site, an autonomous car. (e) Robot reached at its third visit site, AMAL. (f) Suddenly the robot stops because a visitor is not visible and it waits for the visitor to appear in the field of vision. (g) Robot sees the visitors and detects their faces and makes an announcement to move on. (h) Robot moves towards its next visit site, Tinku a social robot for kids. The robot explains interesting facts about Tinku. (i) Robot moves towards its last journey point (j) Robot arrived at its last visit place, SMART (k) The robot terminates its tour successfully completes its robotic journey.

Now the robot stops and explains about the NaO and AmigoBot robots. After explaining everything about the NaO robot and AmigoBot, the robot again started its journey while simultaneously detecting the visitors. Followed by a small-time gap, the robot reached its second visit site where the mini self-driving car was parked. The robot stopped here and introduced the autonomous car to the visitors as shown in Figure 4.6 (d). The robot thereafter reached its third visit site where the Adaptive Modular Active Leg (AMAL) was present and introduced important facts to the visitors in the group and moved to the next goal, shown in Figure 4.6 (e). However, after visiting AMAL, a visitor was not found and therefore the robot stopped here and made an announcement "*looks like I am missing someone, let us wait*" and the robot waited for the missing visitor as shown in Figure 4.6 (f).

As the visitors appeared in the camera mounted on the robot, the robot detected the visitors and made an announcement "*Okay! I have found you let us move*". The robot restarted its delayed and remaining journey. Figure 4.6 (g) shows the robot detecting the visitors and moving towards its fourth visit site and simultaneously visitors are also following the path navigated by the robot. In Figure 4.6 (h), the robot arrived at its fourth visit site where the *Tinku* robot was positioned. The robot stopped here and told interesting facts about the *Tinku* robot to the visitors in the group Now only one place was left in the journey. The robot was ready to departure for its last goal as shown in Figure 4.6 (i). The SMART robot was located at the last visit site in this robotic tour. In Figure 4.6 (j), the robot successfully reached its last visit site. The robot stays here and gives the brief explanation of the SMART robot to the visitors. Followed by all these steps, the robot successfully completed its robotic tour and navigated the visitors, shown in Figure 4.6 (k).

4.7.3 Impact of Behaviour Finite State Social machine on real tour with multiple scenarios

In this section, the results of the proposed methodology have been discussed. This method has been tested on three scenarios. In the first scenario, the robot starts its journey from the starting point, and visits the visiting sites and finally reaches the last visiting point. In this case each person follows the robot and no-one misses the middle journey section. The second scenario is the same as the first scenario but both visitors are missing from the group in two different places during the tour. Moreover, in the third scenario, one visitor misses the tour at one place and both visitors miss the tour at another stage of the tour. The visitors going missing may be due to cause such as receiving a phone call, moving away from the group as the visitor sees something interesting, retrieving wandering children, etc. While any visitor is missing from the group or tour, the robot decreases its speed. Meanwhile, if any visitor re-joins the group, then the robot increases it speed and moves towards the next point of interest. If the visitor does not re-appear, then the robot stops and makes an announcement requesting for the visitor to join the group followed by waiting for the person to appear or re-join the group. As soon as the person re-joins the group, the robot makes an announcement and continues the tour. Detection of visitors takes less than 1 second. Time taken to stop on detecting visitors leaving the group is an important key performance indicator and the robot takes 3 seconds for stopping upon detecting missing visitors which is fixed as an algorithmic parameter. Reducing this parameter risks the number of false positives. Detecting people re-joining the group takes less than 1 seconds.

The proposed approach uses interesting heuristics and the uncertainty-based navigation behaviour of the robot. In the literature no competent approach was found to the same problem that could be used for comparisons. Hence the comparison of the approach is carried out on a baseline. The baseline works on the principle of making the robot move as long as all visitors in the group can be tracked. If the visitors can no longer be tracked, the robot stops and makes an announcement for the missing visitors. Table 4.2 describes the metric comparisons of both the proposed approach and the baseline approach. Three metrics are considered. The first metric is the total time to complete the tour. The robot in both cases does not move while the visitors are missing. Ideally, the robot should only wait for the length of time the visitors are not available. Any additional time waiting hence is the robot waiting as the visitor is not currently visible or the robot making the wrong decision. The metric is modified to account for the time when the person was actually not present as adjudged by a human looking at the scenario. The false positives, or cases when the robot believes the person is not present and makes an announcement, while the person is actually there, however may not be visible to the robot, is another metric. The proposed approach easily outperforms the baseline on all metrics by a major factor. The robot finishes its tour comparatively early with the proposed approach and moreover the false positives are also zero, while in the baseline method the false positives exist which delays the tour. This should be visualised as the visitors waiting and adjusting themselves to be within the view of the robot which can be a painful experience often observed with the adoption of technology in non-ideal circumstances.

Table 4.2: Comparison between the proposed method and baseline

Scenarios	Scenario Description	Time duration of tour (s)		Time duration of tour excluding visitors not available (s)		False Positives	
		Proposed	Baseline	Proposed	Baseline	Proposed	Baseline
1	No visitors leave	**124**	152	**124**	152	**0**	2
2	2 visitors leave at V_1^N and 2 visitors leave at V_4^T	**154**	205	**119**	172	**0**	3
3	1 visitor leaves at V_1^N and 2 visitors leave at V_4^T	**147**	228	**119**	196	**0**	4

At the baseline, some of the times the visitors are present behind the robot however the robot is not able to detect them and stop. When the robot reaches at a corner and makes a sharp turn towards its next visit site, then the robot is not able to detect the visitors and stops and waits for them to appear in front of the camera, while there is not enough space behind of the robot as shown in the following Figure 4.7(a). Waiting makes the problem complicated as the robot will not move until all the visitors are visible, however there is a very little space behind the robot standing at the corner and the visitors cannot fit themselves in this little space. In the proposed approach, this thesis resolved this issue and the robot can easily tackle such turns as shown in Figure 4.7(b), which is innovative and contributes towards a better resulting performance.

The result of the first test case for the first scenario is shown in Figure 4.8. Here every visitor followed the robot and also false positives were zero and the tour completion time was short. In this scenario the robot started its journey at its initial state and completed its journey at its last visiting site. The robot could also tackle sharp turns easily. Different snapshots from the tour are shown in Figure 4.8.

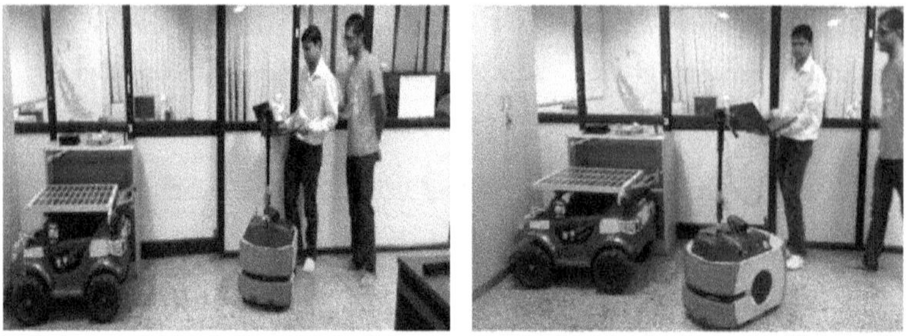

(a) (b)

Figure 4.7: Robot location at a corner

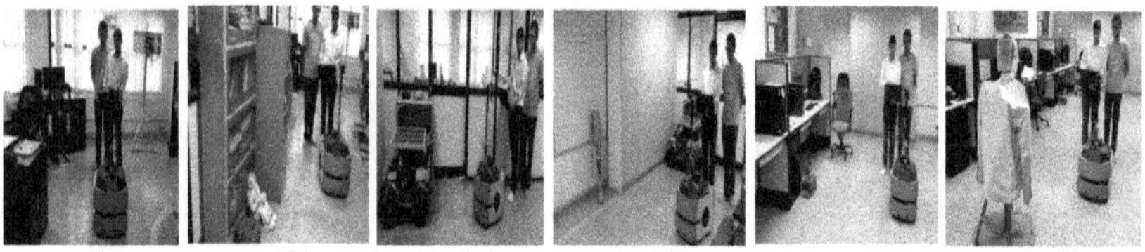

(a) (b) (c) (d) (e) (f)

Figure 4.8: Robotic tour for scenario 1 (a) Robot at its initial state and is ready to start its tour. (b) Robot successfully arrived at its first visiting site where NaO and AmigoBot is placed. (c) Robot at its third visiting site where the autonomous car is parked. (d) Robot successfully made a sharp turn and reached the third visit site, AMAL. (e) Robot at the fourth visite site where Tinku a social robot is present. Meanwhile the robot also took a sharp turn to reach here. (f) Robot reached at its last visit site and SMART robot is located here. Also the robot terminates here.

(a) (b) (c) (d) (e) (f)

(g) (h) (i) (j) (k) (l)

Figure 4.9: Robotic tour (a) Robot at its initial state and is ready to start its tour. (b) Robot successfully arrived at its first visiting site where NaO and AmihgoBot is situated. (c) Here both visitors left the group and not detected by the robot. So, the robot waited for those visitors to join the group. (d) All visitors are detected by the robot. (e) Robot at its third visiting site where the autonomous car is parked. (f) Robot successfully made a sharp turn to proceed to its next visiting site (g) Robot successfully reached at the third state, AMAL. (h) Robot again made a sharp turn at the corner to proceed its next visiting point. (i) Robot at the fourth visite site where Tinku a social robot is present meanwhile robot. (j) All the visitors are not detected by the robot behind it so the robot waited for the visitors. (k) Visitors appeared behind the robot and were detected by the robot (l) Robot reached at its last visit point and SMART robot was located here. The robot terminated.

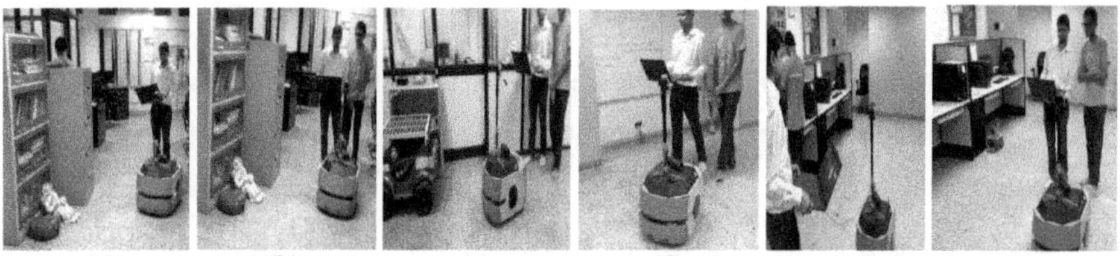

(a) (b) (c) (d) (e) (f)

Figure 4.10: Robotic tour (a) Here only one visitor left the group while the other visitor was available, the robot still waited. (b) All visitors are detected by the robot and it resumed its journey. (c) Robot succeessfully made a sharp turn at the corner to proceed to its next visit site. (d) Robot again made a sharp turn. (e) All the visitors moved away in different directions and were no longer "seen" by the robot, so the robot waited for them. (f) All the visitors came near the robot in the rear side and are detected by the robot, so the robot moved on for its next visit site.

(a) (b) (c) (d) (e)

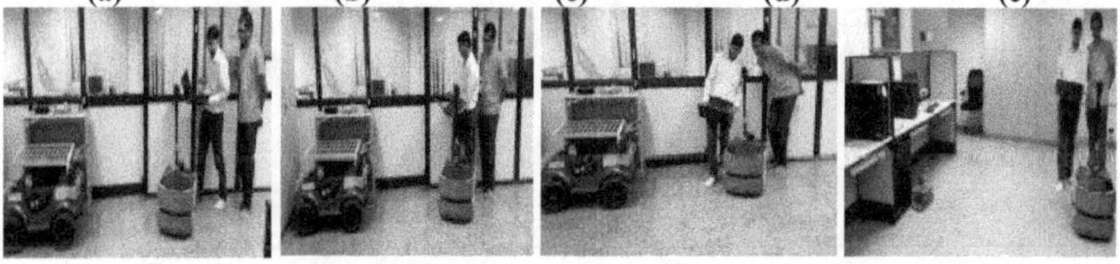

(f) (g) (h) (i)

Figure 4.11: Robotic tour for the baseline (a) Robot suddenly took a 90° rotation and produced false positives while both visitors were present behind the robot. (b) Robot took a turn at a corner in such a way there was not enough space behind the robot. (c) Visitors tried to make themselves seen behind the robot however the robot did not detect them. (d) Visitors try to place themselves within the field of view of the camera, but still could not be detected. (e) Robot took a turn with 90° rotation and generated false positives. (f) When the robot was taking a turn there was no sufficient space behind the robot and that produced false positives. (g) Visitors tried to appear behind the robot. (h) Visitors appeared behind the robot but they were not detected by the robot, so changed their face orientation in accordance with the to view of the camera mounted on the robot. (i) The robot took a 90° rotation to reach its next visit site and also produced false positives even though visitors were available behind the robot.

Further, in the second test case tour, the robot navigated all the visitors and also made sharp turns at the corner very easily. In this scenario visitors were also following the robot somewhere in the middle, so the robot decelerated and stopped, and made an announcement to wait for them. As the visitors joined the group and appeared behind the robot, the robot started its tour. All these cases are shown in the Figure 4.9.

The third test case was similar to the second one. The only difference was that at one state only one visitor left the group. So, the robot waited for that visitor. Some specific snapshots of this tour are shown in Figure 4.10.

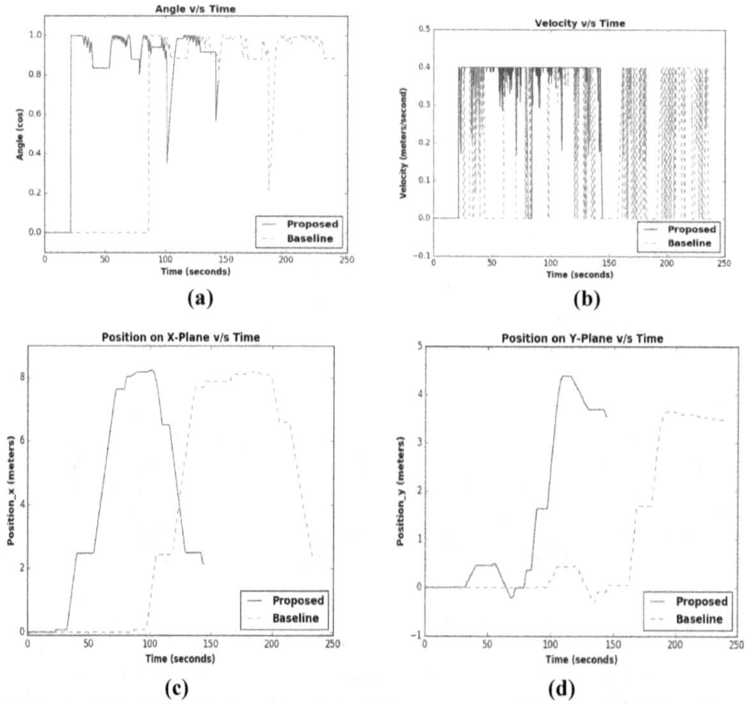

Figure 4.12: Test case I comparison between proposed method and baseline

Similarly, the tour is discussed for one case for the baseline algorithm. False positives existed here, and the robot stopped while the visitors were present behind the robot and were following the robot. A major problem arose when the robot reached the corner and could not move as the visitors were not visible. While it was making the turn with this method, then it suddenly took 90° rotation at the corner. In these circumstances, the visitors could not follow the robot. Some of the time, there was not enough space behind the robot at the corner where visitors could place themselves. Even when the visitors were trying to place, their face was not easily detected and the view of the camera mounted on the robot had to be changed manually to look behind. These situations are shown in Figure 4.11.

At different times in the duration of the robot navigation, angle to goal (rotation, given by the cosine value), and velocity of the robot have been determined for each case. The graphic representations these trajectory, angle, and velocity of scenario 1, scenario 2 and scenario 3 are shown in Figure 4.12, Figure 4.13, Figure 4.14 respectively. All the plots the in blue colour are for the proposed approach and the red colour is for the baseline.

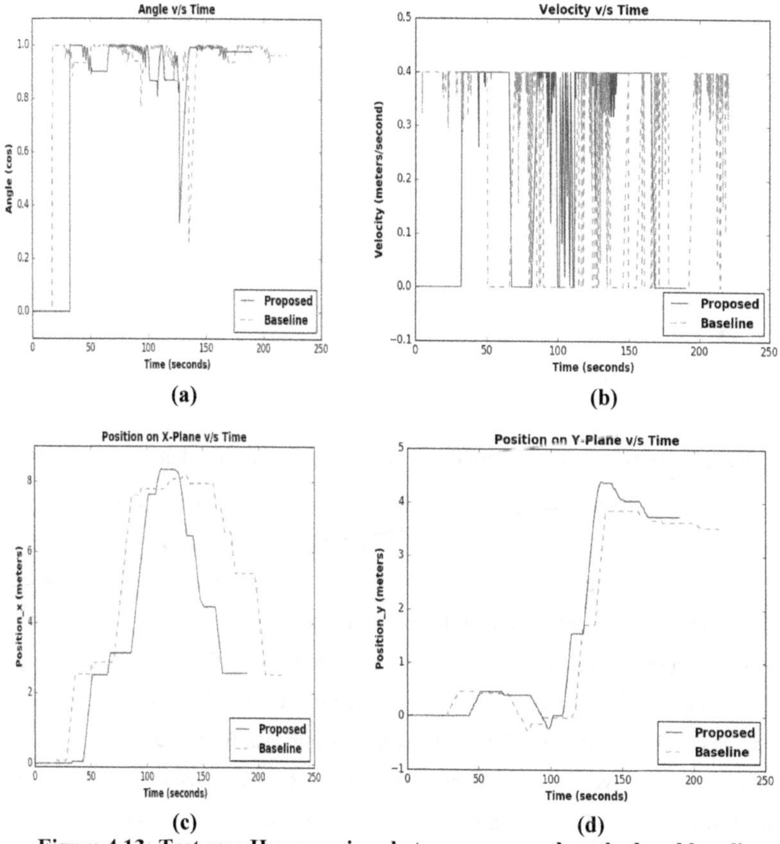

Figure 4.13: Test case II comparison between proposed method and baseline

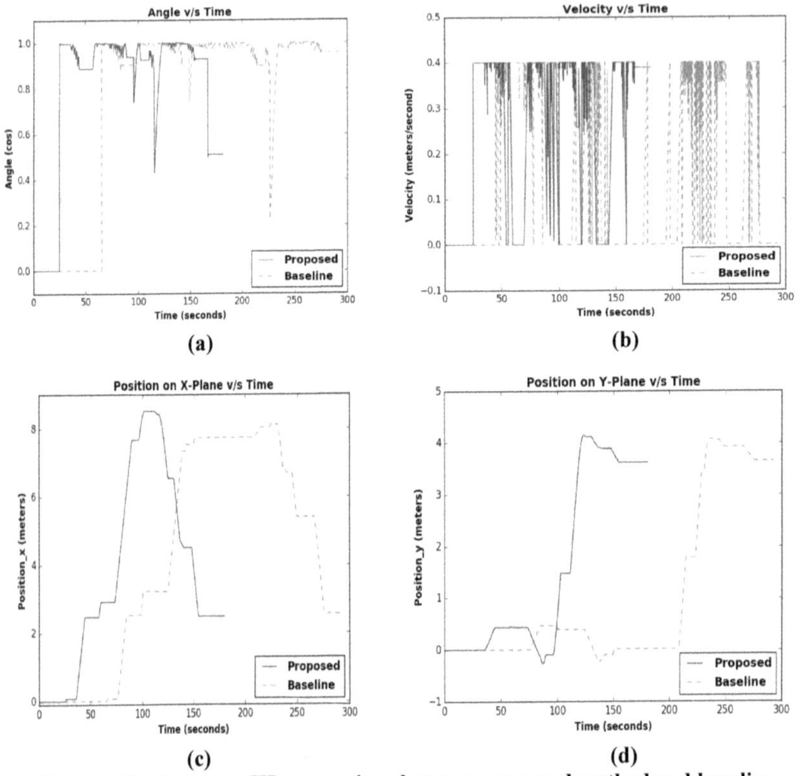

Figure 4.14: Test case III comparison between proposed method and baseline

With these metrics of comparison between the proposed approach and the baseline this thesis state that the proposed method takes less time for the tour completion as compared to the baseline. Moreover, the speed of the robot is disproportional to the time the visitors were missing. If the visitors are available at the front of the camera mounted on the robot, then the speed of the robot will be at a maximum defined by the speed controller and if the visitors are not present behind the robot then the speed of the robot will decrease to zero. Therefore, these are distinct parameters to evaluate the performance of the proposed approach work and this thesis found that this proposed method gives the best performance in terms of journey completion time, with the robot slowly decreasing and increasing velocity. The robot does not become stuck at a corner while taking a turn, and also maintains a sufficient space so that the visitors can easily follow the robot and be detected by the robot.

4.8 Conclusion

This chapter presents a social robot that can take a group of visitors on a tour of a facility. Initially experiments were carried out assuming that one module could be used to sequentially move the robots to different visit sites, stopping and explaining the different visit sites. A second module could then be used to detect visitors and give social gestures using the audio channel. However, two problems were faced. First it was observed that as the robot reaches one visit site and explains the important facts of that visit site; at the same time if a visitor gets missing then the robot also must make an announcement asking the missing visitor to join the group. The robot announced both things simultaneously which was erroneous. Therefore, an architecture level modelling was carried out so that the two modules do not simultaneously make an accouchement. The second challenge was that if a visitor wanted to take longer time at a site or was lagging for any reason, the robot would continue the tour at the same pace that was not socially pleasing. Hence a challenge was to integrate the two modules to make the robot stop if necessary.

This chapter produces two different finite state machines for the social robot navigation. The first machine is named as 'Behavioural Stop & Wait Machine' that enables a robot to move towards the goal and stop when the visitors are not detected by the robot. Initially, visitors are detected by their faces. A static map is loaded on the robot. The robot can compute its current pose by a localisation module. The robot is made to sequentially visit different sites that are placed on the map with their 2D coordinate points. BSW machine uses the ROSARNL library for motion planning. The robot continues the tour only when the visitors are available.

A major problem with the BSW machine is that the robot suddenly stops if the visitors are lagging that is not socially pleasing. The design was thus modified to let the robot slow down as a social gesture to give more time for the person to catch up. This increases another module that continuously tracks the situation and manages the speed of the robot. The module interacts with the low-level motion planning module and the sequencing module. Letting the three modules run in-parallel but allowing them to continuously give feedback to each other using communication protocols is an application design challenge demonstrated in the thesis.

The second finite state machine is named as 'Behavioural Finite State Social Machine' that uses artificial potential approach for the motion planning. BFSSM consists of several states and transition conditions to move from one state to the other. Initially, the robot is at a 'moving' state when all the visitors are detected. If some visitors are not detected for some time duration, then the robot will go in the 'stop' state and stop. Followed by this the robot makes an announcement for missing visitors to re-join the group and robot will go to the 'announce wait for visitors' state. The state of the robot will again change if the missing visitors are found. Here the visitors are detected by their faces, but the robot does not stop suddenly if visitors are not available behind the robot. BFSSM computes the time duration for which the visitors are not detected. When visitors are not available for a certain time period then only the robot goes in the 'stop' state and stops. When visitors are not detected for a small time, then the robot gradually decreases its velocity to give a social signal to the visitors.

BSW machine produces false positives (assumes visitors are missing while the visitors are available behind the robot). BFSSM produced 'no false positives' in the tested scenarios, which is a major achievement of the work. Due to higher false positives BSW machine takes a long time to complete the journey. BFSSM knows the uncertainty about the visitors at the corner. When the robot takes a turn at a corner then due to the limited field of view of the camera and a lack of space the visitors are not detected by the robot. In BFSSM the robot will neither stop nor produce false positives. But these uncertainties are not inbuilt in the BSW machine. When a robot uses a BFSSM, then the robot does not stop suddenly even if the detection of the face losses.

The assessment was carried out in three diverse scenarios. In the first scenario all visitors ideally follow the robot and do not miss the group. The proposed method takes 124 seconds to complete the tour whereas the baseline approach takes 152 second and it also produces 2 false positives. Further, two more different scenarios were used to check the performance of the proposed method where visitors are missing the group at different places. Here the robot takes 154 seconds (scenario 2) and 147 seconds (scenario 3) to complete the tour whereas the baseline approach takes 205 seconds (scenario 2) and 228 seconds (scenario 3) and also produces 3 (scenario 2) and 4 (scenario 3) false positives. If the time duration of missing visitors is excluded, the journey is completed in 119 seconds in both scenario with the proposed method, but the baseline approach takes 172 seconds (scenario 2) and 196 seconds (scenario 3). The tour times are reduced as compared to baseline approach by 18.43% (scenario 1), 24.85% (scenario 2), and 35.53% (scenario 3) if the tour time metric includes the missing visitors' time; and 18.43%

(scenario 1), 30.82% (scenario 2) and 39.29% (scenario 3) if the tour time metric excludes the visitor missing time.

In the baseline approach if the face is not detected, the robot stops within a second. But in the proposed method the robot detects the humans' availability according to time duration of visibility. If a visitor is not detected for upto 3 seconds, the robot will slow down its speed gradually. It may appear that in case of missing visitors the baseline's immediate stopping is better; however, when the visitors are available but not detected, false positives occur that harms the algorithm more than the additional time of 3 seconds used by the BFSSM in slowing.

Currently the methods were tested in a small area (7 meters long and 4 meters wide) that takes already less time (2 to 3 minutes). The baseline method generates a small number of false positives (2 to 4) while saving 30-80 seconds. These number of false positives generated by the baseline and the saved journey time may appear very small. But when the tour executes on a large area typically for over an hour, then the number of false positives will be significantly large for the baseline. As the false positives will increase, the journey time will also increase. All the visitors come to a tour to enjoy. If there are many false positives, for every false positive every visitor must look into the camera so that the robot can detect them. Visitors cannot enjoy the tour in such a situation and their full attention will be towards the camera only that will be a displeasing experience, encouraging visitors to terminate the tour. There is a greater risk associated with the baseline approach. At a corner when the robot makes a sharp turn, it may have a wall at its rear with a very limited space. The robot will not move until it sees all the visitors while the visitors cannot place themselves behind the robot. In this case, the robot will be in the waiting state indefinitely and will not move further.

There are a few important limitations of the approach. The current approach reaches visit sites by navigating through multiple sub-goals. Many times, environments get congested with tourists requiring the guides to appropriately re-route the group, a capability absent in the current mechanism. Many times, the current sub-goal may not be achievable due to a changed map that the system currently cannot mend. If there are multiple obstacles in the workspace (dynamic as well as static), so due to limitation of space in the workspace, a robot may not be able to predict the path of each visitor. Guides often ask the group to directly meet at the next site in such situations. This also needs to be accounted for in the algorithm. Currently there is no time limitation to complete the tour. Practically tour guides often adjust the time to spend at each site and the list of sites also varies as per the time limitations, which is not incorporated in the current approach. The proposed scheme is for small group tours only. The face detection algorithm has a large scalability and will continue to have a reasonable response time that does not limit the algorithm's performance. However, in large groups, it is impossible for all the visitors to come inside the same image. In tours, visitors typically get close to the guide and in the case that the group is large, the visitors spread all around. The limited field of view camera can thus only see a sub-set of visitors. This is an actual problem with tour guides in large groups as well, wherein counting the visitors only takes place at major stops by actually roaming around the group before departing.

Chapter 5

Social Robot Navigation

5.1 Introduction

Due to advancements in robotic technologies in recent years, there has been a lot of research in the area of social robotics (Breazeal et al., 2016; Alač et al., 2011), where robots work alongside humans. Such services include room service, luggage carrying services in areas such as shopping centre, railway stations and airports (Weiss et al., 2014; Tsuda et al., 2009; Shiomi et al., 2009). These tasks are very complex because human behaviour is unpredictable.

5.1.1 Social Cues

The humans usually follow a few subtle behaviours known as social cues for easy navigation in crowded environments. Social cues are usually non-verbal hints that can be used to guide interactions and help to clarify people's meanings and intentions. It is important that the robots exhibit social cues to be recognised by the humans and the social cues be like those used by the humans.

While navigating an environment with many humans around, they maintain a reasonable distance from the others passing by, and the robots must do the same (Bera et al., 2017; Burgoon and Jones, 1976). It has been widely observed that the humans follow simple navigation rules to avoid colliding with each other such as passing on the left or overtaking on the right (Pandey and Alami, 2010; Kala and Warwick, 2013). These rules are derived from the traffic conventions using a left-handed convention, which may be reversed for a right-handed convention.

The first social cue introduced is passing on the left. If two humans walk towards each other head on, it is evident that both will drift leftwards to avoid each other. The specific behaviour is inspired by the traffic (Kala and Warwick, 2013). In countries where traffic is unorganised and lane discipline is not mandatory (e.g., India), closely representing a mobile robotics problem, it is common for two vehicles to come head-on towards each other. However, both vehicles drift sideways in anticipation to avoid a collision. The robot must similarly drift leftward on encountering a person directly ahead. The second social cue is enabling overtaking on the right. If a human is getting late and walking fast, with a slow-moving human ahead, an overtaking takes place, if the two humans are oriented in the same direction. The overtaking human drifts rightwards, while the human being overtaken drifts a little leftwards.

The human-in-the-loop is a very recent convention in robotics, wherein the robot is made to operate in the presence of humans (Gopinath et al., 2017; Schirner et al., 2013). The field of social robot motion planning (Ferrer et al., 2013; Sisbot et al., 2007) enables making the robots operate in workspaces in the presence of humans, following social behaviour while making the overall motion collision-free. The current approaches are limited to human robot interaction, manipulation in cooperation with humans, extrapolating the human trajectory to avoid them, etc. There has been a very limited literature towards the navigation of robots in such contexts. The navigation literature is also limited to the maintenance of social distances, which finds a lot of application in simulating crowd behaviour. This chapter proposes to solve the problem of social robot motion planning. This thesis first observes the generic behaviour that the humans involuntarily display while navigating in home or office environments. Thereafter, the robot navigation strategy is designed such that these social cues are anticipated as well as displayed by the robot.

5.1.2 A brief of the approach

There are two dominant techniques for the reactive navigation of the robot. The first is the social potential field (Helbing and Molnar, 1995; Reif and Wang, 1999). Here, every obstacle applies a repulsive force, while the robot experiences an attractive force towards the goal (or in the direction of motion). The method is social because of the social modelling of the forces by the humans. This thesis extends the social forces model by introducing additional forces to enforce the conventions to be followed by the robot.

The second dominant approach for the reactive navigation of the humans involves a geometric assessment of the obstacles (Sezer and Gokasan, 2012). Typically, the obstacles are assessed to get the gaps, within which the robot can place itself. Out of all the gaps, the best one is chosen, and the heading direction is kept to avoid the obstacles with a safe distance, while keeping generally towards the goal. This thesis extends the geometric approach by introducing a two-stage gap qualification and gap-selection strategy, and then introduces social cues into the selection strategy in an intuitive manner.

Both potential field and geometric approaches have contrasting pros and cons which motivate the fusion of the two techniques making a hybrid reactive algorithm for the social navigation of the robot (Kala, 2018; Murphy, 2000). This thesis designs a novel hybrid method by combining the force model with the geometric approach in a way that both their characteristics are retained and the social cues are displayed. This thesis demonstrates that the hybrid method generates safer trajectories and maintains a larger clearance from the humans.

The key performance indicators are:

i. Minimising the penalty for violation of social distance between the robot and humans. The minimum social distance is computed from a recorded human trajectory database.
ii. Ability of the algorithm to imitate human trajectory available in the human trajectory database.
iii. Adherence of the social norms of overtaking on the right and passing from the left.
iv. Clearance
v. Travel time

5.1.3 Main Contributions

This chapter assumes a general social convention that the humans prefer to avoid each other from the left and allow people to overtake by drifting in the other side. The novelty of the work is adapting the current reactive motion planning algorithms to include this social behaviour in their working. It is one of the first approaches that directly translates social human behaviour into the existing motion planning methods from robotics literature. The main contributions are:

- Fusion of a geometric approach which finds the best social gap and a potential social approach for avoiding colliding with humans.
- Adding a new social force in the potential field approach orthogonal to the vector to the human. The selection of leftward or rightward orthogonal direction vector is carried out based on the social preference.
- Using social cues to select a gap for the navigation of the robot, to make the resultant approach socially compliant.
- The results are demonstrated under live settings using the Pioneer LX robot.

Based on the experimental results, the hybrid approach performs better than the baseline and individual approaches.

5.1.4 Organization

This chapter is organized as follows. Section 5.2 summarises the related works carried out in the literature. Section 5.3 explains the social potential field algorithm and the social cues that are induced in the social potential field algorithm. Section 5.4 provides a description of the gap-based strategy. The section also introduces the social cues in the gap-based strategy. Further, a Hybrid force model and its algorithm is explained in section 5.5. Section 5.6 describes the implementation details and different computations such as face distance and angle. Experiment evaluation and result analysis of each algorithm is discussed in section 5.7. Section 5.8 gives the conclusions of this chapter.

5.2 Related Work

The existing literatures on decentralised, cooperative and socially compliant navigation can be broadly divided into either model-based approaches or learning-based approaches. Model based approaches are generally extensions of multi-agent collision avoidance algorithms by introducing additional parameters for social interactions. It has been widely observed that humans follow certain conventions such as passing on the left and overtaking on the right according to the left-hand convention present in the UK, India etc. (Knepper and Rus, 2012). The same can be reversed such as passing on the right-hand convention present in the US, Japan, etc.

5.2.1 Geometry-based Approaches

The classical planning algorithms can be used to generate a collision free path for the robot under the assumption that the human's trajectory is unknown. It has however been observed that separating navigation into independent prediction and planning leads to freezing robot problem. The robot in such cases fails to find a path as a major part of its environment is marked as untraversable. One of the most common approaches to the problem of collision avoidance with dynamic obstacles is to employ specific reactive rules in order to avoid collisions. Reciprocal velocity obstacle avoidance algorithms implicitly assume the behaviours of other agents to generate collision free trajectories (Van et al., 2008; Van et al., 2011; Zhang et al., 2015). Since these methods do not capture human behaviour, they can sometimes result in generating dangerous trajectories. An important factor is to account for cooperative mechanisms between the robot and the humans to make navigation easier. Cooperative mechanisms help to make navigation possible in some situations that could previously cause the freezing robot problem. An adaptive side-by-side navigation methodology has also been also introduced to guide the robot (Repiso et al., 2019). The approaches assume a well-known noise-free robot position and velocity, with a circular geometry. The approaches do not account for static obstacles of arbitrary shapes and rely on a deliberative technique to handle the static obstacles. This limits the performance when the reactive motion leads to a position contradictory to the deliberative plan.

The work of Pandey and Alami (2010) is conceptually closer to the proposed work. The authors modelled multiple rules along with a deliberative planning of the robot. The proposed approach is purely reactive in nature, which is more suited when multiple people walk around in the proximity, nullifying most of the initial deliberative planning. Pandey and Alami (2010) used a milestone-based approach. The motive of this thesis is the fusion of the milestone-based approach with a social potential field approach, which by intuition and experiment both produce better results than either of the social techniques. Unlike Pandey and Alami (2010), this thesis uses the constants of potential field, geometric approach, and controllers to reproduce the behaviour and only externally fuse a single behaviours of social person avoidance. This method eliminates the problem of having an ensemble of rules that are hard to maintain. It also avoids the problems of disagreement between rules when many of them are simultaneously valid. The method does not add selection parameters, while having an ability to demonstrate most of the behaviours.

Geometric approaches try to formulate a solution to the problem by finding the safest portion of the environment for the robot to traverse. The Follow the Gap method is a safety-first collision avoidance algorithm which works on the principle of finding the largest gap in its surroundings and moving through that gap (Sezer and Gokasan, 2012). The algorithm is very attractive as it requires very little parameter tuning. The improved follow the gap method aims to fix one of the primary limitations of the follow the gap method which is that the extremely long trajectories often fail to reach the destination (Demir and Sezer, 2017; Özdemir and Sezer, 2018). One of the major limitations of the geometric approaches, which is an oscillatory behaviour for very small changes in direction, remains. Vector Field Histogram (VFH) is another geometric approach and works based on calculating the obstacle density in the robot's surroundings (Borenstein and Koren, 1991). The environment is divided into different sectors whose densities are found out and the sector with the least obstacle density is chosen to pass through by the robot.

The VHF$^+$ method attempts to account for the size of the robot and the VHF* method aims to combine A* search algorithm in the search space to find the optimal path (Ulrich and Borenstein, 1998; Ulrich and Borenstein, 2000). Search algorithms are however time consuming and expensive in finding acceptable paths in environments containing humans. The approach models all entities as static obstacles and does not account for social interactions.

5.2.2 Potential-based Approaches

Artificial Potential Fields (APF) are a very popular class of techniques used for collision avoidance in static environments (Helbing and Molnar, 1995; Reif and Wang, 1999; Khatib, 1985). The potential field algorithms work on the idea that a virtual force field exists in the environment around the robot. The virtual force field is made up of a combination of attractive and repulsive forces applied by the surroundings. The goal/destination applies an attractive force to the robot while the obstacles apply a repulsive force to the robot. These force models can be extended to account for social interactions by introducing additional parameters which come into play in the presence of humans (Ferrer et al., 2013; Ferrer et al., 2013). The social parameters are determined by running multiple real-life trials consisting of participants (Zanlungo et al., 2011; Shiomi et al., 2014).

The overall problem in this chapter is the same as that found in the work of (Takahashi et al. 2009). However, while Takahashi et al. (2009) modelled the persons as obstacles with different force constants, the approach presented in this chapter uses a very different social force modelling. Here a force is applied orthogonal to the direction of the person. Here the general observation of the proposed approach in social potential field-based methods is that the robot gets too close to the person when the person moves ahead suddenly. Therefore, this thesis models an additional orthogonal force that quickly de-aligns the robot when the person and robot are directly ahead of each other. The selection of avoidance from the left or right direction is another major challenge taken in the proposed approach.

Görür and Erkmen (2017) solved a related problem of social interaction between a person and the robot, where the robot was required to get attention of the person. The authors predicted the person's trajectory and used the same for reshaping the trajectory separately in cases when the person's behaviour was found confident or suspicious. The proposed approach models the avoidance behaviour in corridor like scenarios where the human's trajectory is easy to predict.

Furthermore, in this thesis people are always confident and dominate behaviour and the robot reciprocates. This thesis socially models the robot's navigation behaviour under these assumptions.

Another set of approaches aim to integrate two or more than two collision avoidance strategies into a single model, such that the model retains the advantages of both the models. This fusion is carried out such that the advantages offered by the two techniques complement each other and do not work against each other. Ideally, a proactive and a reactive approach can be combined into a single model so that it can have the advantages of both the models in the resultant model. The extended forces model is used in conjunction with the reciprocal velocity obstacles algorithm where the former is used to deal with the navigation aspect of the problem while the latter deals with the human interactions and collision avoidance (Truong and Ngo, 2017). A modified social forces model considered the body pose and face orientation of the pedestrians in the surroundings to avoid collisions in a more human like manner (Ratsamee et al., 2013).

5.2.3 Other Approaches

Learning based techniques aim to learn to emulate human behaviour from the human trajectory data and try to match the metrics such as socially acceptable distances (Luber et al., 2012). The supervised learning approaches attempt to collect large amounts of data through teleoperation and then train a supervised training model using the collected data (Long et al., 2017; Khare et al., 2018). Deep reinforcement learning is used to train an agent to learn to navigate safely and avoid collisions in an unknown environment (Chen et al., 2017; Everett et. al., 2018). The reward function used to train a reinforcement learning model can be carefully designed to incorporate the social cues into the navigation mechanism by penalising the non-desirable behaviours (Chen et al., 2017). Inverse reinforcement learning is a technique that has been applied to learn a cost function from the human trajectory data and use it to train a reinforcement learning model (Abbeel and Ng, 2004; Ng and Russell, 2000; Kim and Pineau, 2015; Kretzschmar et al., 2016). Autonomous navigation can also be carried out for a human conformability safety framework (Truong and Ngo, 2016). In addition, an offline path planning algorithm has been developed which is goal oriented and close to the bug algorithms (Paliwal and Kala, 2018).

Some other algorithms from different schools of learning are described in the survey paper by (Kruse et al., 2013). In general, the aim of this thesis is not to follow a deliberative path, but to use only a few sub-goals. This is because humans will cause the robot to make a large deviation.

In this application further, most humans suddenly move ahead of the robot or move opposite to the robot.

Based on the literature, it is evident that the investigation of social cues and their incorporation directly into the model-based approaches is currently absent. Even though methods have been used to make the approaches maintain social distances with humans, the larger cues of allowing overtaking on the right and avoiding on the left are absent from the literature. The learning-based approaches are not suited to the problem since the robots being socially and kinematically different from humans cannot simply imitate the exact human behaviour. The learning of human behaviour is itself a hard task. Learning in the presence of humans and robots is a data and computationally intensive approach.

5.3 Social Potential Fields

The introduction of social cues for socially acceptable navigation can be formulated as extensions of existing decentralised multiagent collision avoidance algorithms. In this section the social cues are introduced for a social potential field approach.

5.3.1 Artificial Potential Field

The extended forces model assumes the presence of a virtual force field in the environment of the robot. The force field is made up of the different forces acting on the robot and consists of attractive forces from the goal, repulsive forces from the obstacles surrounding the robot and social forces from the humans. The robot is assumed to be equipped with a laser scanner whose readings are used by the algorithm to calculate the forces of repulsion. The magnitude of the forces applied on the robot are determined by the robot's location. Mathematically, the forces are modelled as follows; the force applied by the goal is proportional to the distance of the robot from the goal while the repulsive forces are inversely proportional to the square of the distance. The social forces accounting for social cues are dependent on the positions of the people in the environment within the vicinity of the robot.

Assume $q(q_x,q_y)$ is the position of the robot and $g(g_x,g_y)$ is the position of the goal. The attraction force is proportional to the distance to goal $d(q,g)$ and is directed towards the goal denoted as a unit vector in the direction g-q, or $\hat{\mathrm{u}}(g-q)$. The distance is given a threshold of d_{max} to avoid excessively large values. The attraction is thus given by Equation (5.1).

$$F_{attr} = \begin{cases} k_{att} d(q,g)\hat{u}(g-q) & \text{if } d(q,g) \le d_{max} \\ k_{att} d_{max} \hat{u}(g-q) & \text{if } d(q,g) > d_{max} \end{cases} \tag{5.1}$$

k_{att} is the attraction proportionality constant. Consider θ_g be the angle to goal, shown by Equation (5.2).

$$\theta_g = \text{atan2}(g_y - q_y, g_x - q_x) \tag{5.2}$$

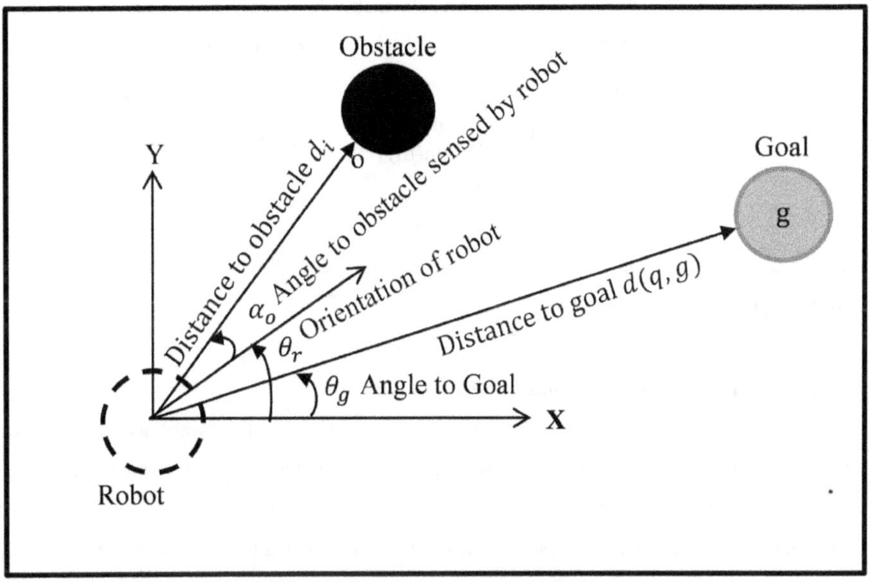

Figure 5.1: Notations used for the calculation of attractive and repulsive forces. The attraction force is proportional to distance to goal (d_g) with an upper threshold, in the direction of goal (θ_g). The repulsive force is inversely proportional to the distance to obstacle (d_o), opposite to the direction to obstacle (α_o) with respect to heading direction of robot (θ_r)

This gives $\hat{u}(g - q)$ for a planar scenario as $[\cos(\theta_g)\ \sin(\theta_g)]^T$. The attraction force acting on the robot is thus given by Equation (5.3). The notations are shown in Figure 5.1.

$$F_{attr} = \begin{cases} k_{att} d(q, g) \begin{bmatrix} \cos(\theta_g) \\ \sin(\theta_g) \end{bmatrix} & \text{if } d(q, g) \leq d_{max} \\ k_{att} d_{max} \begin{bmatrix} \cos(\theta_g) \\ \sin(\theta_g) \end{bmatrix} & \text{if } d(q, g) > d_{max} \end{cases} \tag{5.3}$$

The repulsion force is applied by every obstacle *o*. The repulsive force is inversely proportional to the square of the distance *d(q,o)* and is directed away from the goal, given by the unit vector $-\hat{u}(o - q)$. The repulsive force is given by Equation (5.4).

$$F_{rep}(o) = -k_{rep} \frac{1}{d^2(q,o)} \hat{u}(o - q) \tag{5.4}$$

Here k_{rep} is the repulsive force proportionality constant. It is assumed that the robot has a laser scanner and the ith laser hits at an angle of α_i from the heading direction of the robot (θ_r) to indicate a point obstacle at a distance d_i. The direction of the force is thus $[\cos(\alpha_i)\ \sin(\alpha_i)]^T$ in the robot's coordinate frame. If $T(R,W)$ is the transformation matrix between the robot and the world, the direction of force is given by $T(R,W)[\cos(\alpha_i)\ \sin(\alpha_i)]^T$. For this planar case, transform is a rotation along Z axis or $R_Z(\theta_r)$, and the force is given by Equation (5.5)

$$F_{rep}(i) = -k_{rep}\frac{1}{d_i^2}T(R,W)\begin{bmatrix}\cos(\alpha_i)\\ \sin(\alpha_i)\end{bmatrix} = -k_{rep}\frac{1}{d_i^2}R_Z(\theta_r)\begin{bmatrix}\cos(\alpha_i)\\ \sin(\alpha_i)\end{bmatrix} \tag{5.5}$$

$$= -k_{rep}\frac{1}{d_i^2}\begin{bmatrix}\cos(\theta_r+\alpha_i)\\ \sin(\theta_r+\alpha_i)\end{bmatrix}$$

The resultant repulsive force is the addition of repulsions of all obstacles detected by the laser scanner, given by Equation (5.6)

$$F_{rep} = \sum_i F_{rep}(i) \tag{5.6}$$

5.3.2 Inducing Social Cues in Social Potential Field

Two people walking towards each other prefer to drift leftward as per the social conventions in the author's country, and the same social etiquette must be displayed by the robots. The social force in the extended force model picks a direction to avoid a collision well in advance and the human is expected to reciprocate the behaviour by picking the opposite direction in which to move, thus cooperating to avoid a collision. Overtaking is another behaviour, while common to humans is uncommon to mobile robots because they usually move at much lesser speed than the human walking speed. This results in a consequent behaviour of being overtaken modelled in the robot. If a human is detected behind the robot, then a similar social force comes into play allowing the human to overtake. The robot drifts to one side to enable easy overtaking by the human from the other side.

Humans generally tend to follow a convention when it comes to cooperative collision avoidance strategies. Passing on the left is followed according to the British convention and the same rule can be reversed according to the American convention. When a human is detected walking towards the robot, the social force model applies a force towards the left side, which is inversely proportional to the distance of the human from the robot and the robot picks the left side to pass the human. Similarly, in presence of a human walking towards the robot from behind, a similar force is applied to the left side allowing the human to overtake from the right-hand side.

The problem of oscillatory behaviour is therefore removed with the introduction of the social cues and safer distances are maintained from the pedestrians to not invade their personal space.

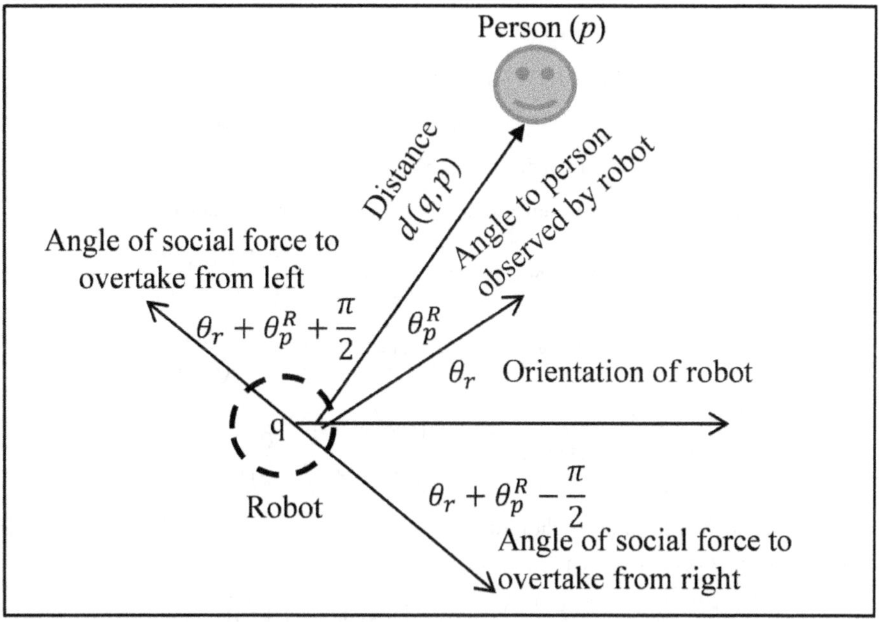

Figure 5.2: Notations used for the social forces. The social force by a person is inversely proportional to square of distance, and is directed in either of the angles shown, orthogonal to the vector to person.

Consider a person p at a distance of $d(q,p)$ from the robot at position q. The social force by the person p is inversely proportional to the square of the distance between the robot and person $d(q,p)$. The force is applied orthogonal to the direction to person. Unlike in some literature, the proposed approach does not treat the persons as they were moving obstacles. The force due to people as regular obstacles is added opposite to the direction of the obstacle, and additionally, for people a social force is added orthogonal to the direction to the person. This results in the approach moving the person to the left or right, depending upon the socially selected side, at each and every iteration. This force results in enough clearance and safety for the robot.

The direction to person is given by $\hat{u}(p-q)$, and an orthogonal direction (leftward) has an additional rotation of $\pi/2$ in Z axis, or $R_Z\left(-\frac{\pi}{2}\right)\hat{u}(p-q)$. The resultant force is thus given by Equation (5.7).

$$F_{soc}^{L}(p) = -k_{soc}\frac{1}{d^2(q,p)}R_Z\left(-\frac{\pi}{2}\right)\hat{u}(p-q) \tag{5.7}$$

Similar to obstacles, modelling the unit vector $\hat{u}(p-q)$ by the relative angle between person and robot (θ_p^R) and further transforming in the word coordinate frame by multiplication with $T(R,W)$, gives an expression similar to the obstacles shown by Equation (5.8).

$$\begin{aligned} F_{soc}^{L}(p) &= -k_{soc}\frac{1}{d^2(q,p)}R_Z\left(-\frac{\pi}{2}\right)T(R,W)\begin{bmatrix}\cos(\theta_p^R)\\ \sin(\theta_p^R)\end{bmatrix} \\ &= -k_{soc}\frac{1}{d^2(q,p)}R_Z\left(-\frac{\pi}{2}\right)R_Z(\theta_r)\begin{bmatrix}\cos(\theta_p^R)\\ \sin(\theta_p^R)\end{bmatrix} \\ &= -k_{soc}\frac{1}{d^2(q,p)}\begin{bmatrix}\cos\left(\theta_r+\theta_p^R-\frac{\pi}{2}\right)\\ \sin\left(\theta_r+\theta_p^R-\frac{\pi}{2}\right)\end{bmatrix}\end{aligned} \tag{5.8}$$

To avoid the person in a rightward direction, the force vector is rotated by $-\pi/2$ in Z axis instead of $\pi/2$ in Z axis, given by Equation (5.9)

$$F_{soc}^{R}(p) = -k_{soc}\frac{1}{d^2(q,p)}\begin{bmatrix}\cos\left(\theta_r+\theta_p^R+\frac{\pi}{2}\right)\\ \sin\left(\theta_r+\theta_p^R+\frac{\pi}{2}\right)\end{bmatrix} \tag{5.9}$$

The social force is typically applied to the leftward direction. However, the avoidance should be to the right-hand side instead if the person is already sufficiently positioned to the left. The robot cannot go far on the left to avoid a person who is already on the left-hand side.

Let θ_p^R denote the orientation of the person as observed from the robot. If θ_p^R is 0, it means that the robot is facing the person head-on and hence avoidance should be on the left. If θ_p^R is highly positive, it means that the person is on the left and the robot should avoid the person from the right. If is nearly 0 or highly negative, the person should be avoided from the left. A limiting value of θ_{left} is given to choose between the 2 strategies. The resultant force is given by Equation (5.10). The notations are shown in Figure 5.2.

$$F_{soc}(p) = \begin{cases} F_{soc}^{L}(p) & \text{if}\,\theta_p^R \le \theta_{left} \\ F_{soc}^{R}(p) & \text{if}\,\theta_p^R > \theta_{left}\end{cases} \tag{5.10}$$

Algorithm 5.1 - Social Potential Fields with Social Cues

Input: laser scan ($D = \{< d_i, \alpha_i >\}$), front people ($P_F = \{p_f\}$), rear people ($P_R = \{p_r\}$), robot pose $q(x_r, y_r, \theta_r)$, goal(g)
Output: linear speed (v) and angular speed (ω)
Method:

1. $F \leftarrow [0\ 0]^T$
2. For $< d_i, \alpha_i > \in D$ do
   a. $F \leftarrow F - k_{rep} \frac{1}{d_i^2} \begin{bmatrix} \cos(\theta_r + \alpha_i) \\ \sin(\theta_r + \alpha_i) \end{bmatrix}$
3. If$(d(q, g) \leq d_{max})$ then
   a. $F \leftarrow F + k_{att} d(q, g) \begin{bmatrix} \cos(\theta_g) \\ \sin(\theta_g) \end{bmatrix}$
4. else
   a. $F \leftarrow F + k_{att} d_{max} \begin{bmatrix} \cos(\theta_g) \\ \sin(\theta_g) \end{bmatrix}$
5. For $p \in P_{rear} \cup P_{front}$ do
   a. if$\theta_p^R \leq \theta_{left}$ then
      i. $F \leftarrow F - k_{soc} \frac{1}{d^2(q,p)} \begin{bmatrix} \cos\left(\theta_r + \theta_p^R - \frac{\pi}{2}\right) \\ \sin\left(\theta_r + \theta_p^R - \frac{\pi}{2}\right) \end{bmatrix}$
   b. else
      i. $F \leftarrow F - k_{soc} \frac{1}{d^2(q,p)} \begin{bmatrix} \cos\left(\theta_r + \theta_p^R + \frac{\pi}{2}\right) \\ \sin\left(\theta_r + \theta_p^R + \frac{\pi}{2}\right) \end{bmatrix}$
6. $\omega_{pref} = K_v \text{angleDifference}(\text{atan2}(F_y, F_x), \theta_r)$
7. if$\omega_{pref} \leq 0$ then
   a. $\omega \leftarrow \max(\omega_{pref}, -\omega_{max})$
8. else
   a. $\omega \leftarrow \min(\omega_{pref}, \omega_{max})$
9. $v = K_\omega \min(d_{front}(q), d(q, g))$
10. return (v, ω)

The resultant force is the vector addition of all social forces due to all people, given by Equation (5.11),

$$F_{soc} = \sum_p F_{soc}(p) \tag{5.11}$$

The social force model consists of three types of forces: (i) repulsive force from the laser scan which gives the distance readings of the obstacles; (ii) attractive force towards the goal; and (iii) social forces enforcing to the conventions to be followed in the presence of persons. The resultant force is the summation of the attractive, repulsing and social forces given by Equation (5.12).

$$F_{res} = F_{attr} + F_{rep} + F_{soc} \tag{5.12}$$

5.3.3 Overall Approach

The robot aims to move in the direction of the resultant force vector or $\text{atan2}(F_y, F_x)$. The angular error is the difference between the robot's orientation and intended direction of motion. Using a proportionate controller, the angular speed is given by Equation (5.13)

$$\omega_{pref} = K_{\omega}\text{angleDifference}(\text{atan2}(F_y, F_x), \theta_r) \tag{5.13}$$

Here ω_{pref} is the preferred angular velocity. K_{ω} is the controller gain parameter. Applying the limits of $[\omega_{max}, -\omega_{max}]$ in the acceptable range for angular speed gives the speed as Equation (5.14).

$$\omega = \begin{cases} \max(\omega_{pref}, -\omega_{max}) & \text{if } \omega_{pref} \leq 0 \\ \min(\omega_{pref}, \omega_{max}) & \text{if } \omega_{pref} > 0 \end{cases} \tag{5.14}$$

The linear speed is kept so as to avoid head-on collision and not to surpass the goal. Using a proportionate controller, the speed is given by Equation (5.15).

$$v = K_{\omega} \min(d_{front}(q), d(q, g)) \tag{5.15}$$

K_v is the controller gain parameter.

The working of the algorithm is shown by Algorithm 5.1.

5.4 Gap based Strategy

Geometric approaches are a class of collision avoidance algorithms that are based on locating the obstacle free regions in the robot's surroundings and picking the best region to pass through. The Follow the Gap method picks the largest gap in the robot's surroundings and moves towards the gap. This strategy however does not account for the direction of the gap with respect to the direction of goal. It can sometimes lead to long and inefficient trajectories if the largest gap found is facing away from the direction of the goal. Even when the direction towards the goal is accounted for in the gap selection strategy, the method is prone to oscillatory behaviour even for very small changes in direction. The idea is to break down the gap selection strategy into a two-way qualification and selection process where the qualification strategy accounts for the size of the robot and its dynamics while the selection strategy accounts for the social cues.

5.4.1 Computing Gaps

The first aspect is to calculate the corners in the map using a laser scan. For a sensor that reports unbound distances, a corner is defined at a point where the distance abruptly changes on a small change in the sweeping angle. Let d_i and be the current and d_{i-1} be the previous laser reading (similarly next). If the absolute difference, $abs(d_i - d_{i-1})$ with the previous reading is high (similarly next), then i^{th} laser has hit a corner at distance d_i and angle α_i, which represents a vector $d_i[cos(\alpha_i)\ sin(\alpha_i)]^T$ with respect to the robot. Corners are an ordered sequence of all such points, given by Equation (5.16).

$$\text{corners} = \left[\text{d}_\text{i} \begin{bmatrix} \cos \alpha_\text{i} \\ \sin \alpha_\text{i} \end{bmatrix} : \text{abs}(\text{d}_\text{i} - \text{d}_{\text{i}-1}) \geq \epsilon \vee \text{abs}(\text{d}_\text{i} - \text{d}_{\text{i}+1}) \geq \epsilon \vee \text{i} = 1 \ \vee \text{i} = \text{len(D)} \right] \quad (5.16)$$

Here [] is used to denote an ordered list. *len*(*D*) is the size of the laser scan. There are two exceptional cases, which is the first laser reading (i=1) and last laser reading (i=*len*(*D*)) in a laser scan with bounded field of view. They are both taken as corner points. ε is a corner acceptance threshold. Every abrupt change registers two corners at two consecutive laser readings. The corners are shown in Figure 5.3.

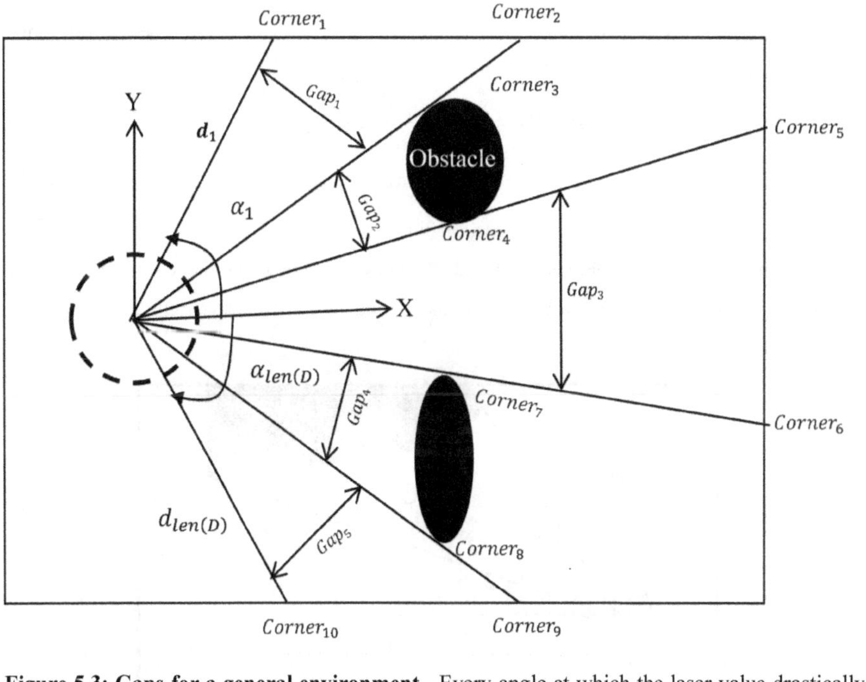

Figure 5.3: Gaps for a general environment. Every angle at which the laser value drastically changes is considered as a corner, including 1st and last laser reading at α_1 and $\alpha_{len(D)}$. Two consecutive corners denote a gap.

The gaps are the open spaces between two consecutive corners. A gap (*f*) is characterised by the two corners (f_{C1} and f_{C2}) with gap as the space between the two corners. The gap array is a collection of all consecutive corners. Since one laser scan gives two corners, the gaps are taken between all odd indexed corner with the next consecutive corner. The set of gaps are given by Equation (5.17).

$$\text{gaps} = \{< f_{C1}, f_{C2} >\} = \{< \text{corners}(i), \text{corners}(i+1) >: i = 1,3,5 \ldots \text{len}(\text{corners}) - 1\} \quad (5.17)$$

The gaps are shown in Figure 5.3. A gap (*f*) is associated with 3 properties. The properties are shown in Figure 5.4. *length*(*f*) denotes the space available to go through the gap. The length is the distance between the corners that make the gap, given by Equation (5.18).

$$\text{length}(f) = d(f_{C1}, f_{C2}), f < f_{C1}, f_{C2} >\in \text{gaps} \quad (5.18)$$

Here distance(*f*) denotes the average distance of the gap from the robot. Every laser angle between the two corners gives a distance, which averaged gives the distance between the robot and gap, given by Equations (5.19-5.21).

$$\text{distance}(f) = \frac{\sum_{i:f_{\theta s} \leq \alpha_i \leq f_{\theta e}} d_i}{\sum_{i:f_{\theta s} \leq \alpha_i \leq f_{\theta e}} 1}, f < f_{C1}, f_{C2} >\in \text{gaps} \quad (5.19)$$

$$f_{\theta s} = \text{atan2}(f_{c1,y}, f_{c1,x}) \quad (5.20)$$

$$f_{\theta e} = \text{atan2}(f_{c2,y}, f_{c2,x}) \quad (5.21)$$

Here $f_{\theta s}$ is the start of the gap angle calculated by the angle to the first corner, and $f_{\theta e}$ is the end of the gap angle calculated by the angle to the second corner. The gap strictly lies between the angles $f_{\theta s}$ and $f_{\theta e}$.

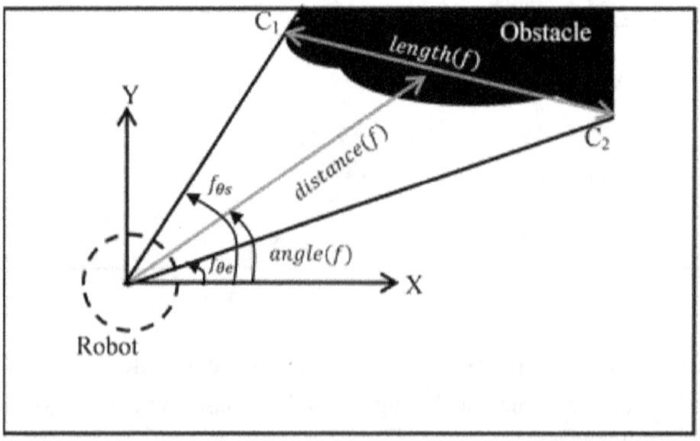

Figure 5.4: Properties of gaps. A gap (*f*), detected by a laser scan between angles $f_{\theta s}$ and $f_{\theta e}$ is attributed by a central angle (*angle*(*f*)), distance to gap (*distance*(*f*)), and length of gap (*length*(*f*)).

The gap mean angle is denoted by *angle(f)* and is given by a weighted average of all angles, weighted by the distance. The formula is similar to the computation of the centre of area. The gap mean angle is given by Equation (5.22).

$$\text{angle(f)} = \frac{\sum_{i:f_{\theta s} \leq \alpha_i \leq f_{\theta e}} d_i \alpha_i}{\sum_{i:f_{\theta s} \leq \alpha_i \leq f_{\theta e}} d_i}, < f_{C1}, f_{C2} > \in \text{gaps} \tag{5.22}$$

5.4.2 Selecting Best Gap

The gaps are filtered by a qualification process to pick those set of gaps which are acceptable ($gaps_Q$). The qualification occurs based on the size of the gap and the size should be at least twice the radius of the robot for it to be an acceptable gap. This enables the robot to pass through. Further, gaps too close to the robot are undesirable and act as obstacles. The qualification criterion of gaps is hence given by Equation (5.23);

$$\text{gaps}_Q = \{f \in \text{gaps} : \text{length(f)} \geq 2R \wedge \text{distance(f)} \geq \eta\} \tag{5.23}$$

Here R is the radius of the robot and η is the gap acceptance threshold. The equation considers all gaps in the gap set, and makes a new set *gapsQ* which have a length of at least $2R$ and are at a distance of at least η.

After the qualification process, the final gap selected is on the basis of being closer to the direction of goal. The gap is at *angle(f)* with respect to the robot, which is oriented at θ_r with respect to the world. The direction of the gap with respect to the world is thus at *angle(f)*+θ_r, while the direction to goal is θ_g. The 'angular error' between the gap and goal is thus *angleDifference(angle(f)*+θ_r - θ_g). The gap with the least 'angular error' is used. The best gap (f_2) is thus given by Equation (5.24).

$$f_1 = \underset{f \in gaps_Q}{\arg\min}\, \text{angleDifference}\big(\text{angle(f)} + \theta_r - \theta_g\big) \tag{5.24}$$

Here θ_r is the orientation of the robot and is added to convert the gaps in the robot frame of reference to the world frame of reference. θ_g is the angle to goal.

The selection strategy is improved with the introduction of the social cues in an intuitive manner. If there are two equally likely gaps on both sides of the goal, then the gap chosen will be the one on the left-hand side. However, if the nearest person is already sufficiently on the left, similar to the potential field, the robot does not try to avoid them from the left side. In this way, the oscillatory behaviour is avoided with the help of the social cues and the extra parameter required for calculating the heading angle in the follow the gap method is removed.

Let f_2 be the 2nd best gap given by Equation (5.25);

$$f_2 = \underset{f \in gaps_Q, f \neq f_1}{\arg\min} \ angleDifference(angle(f) + \theta_r - \theta_g) \tag{5.25}$$

The selected gap is hence given by Equation (5.26).

$$f_{soc} = \begin{cases} left(f_1, f_2) & if abs(angle(f_1) - angle(f_2)) \leq \nu \wedge \theta_p^R \leq \theta_{left} \\ right(f_1, f_2) & if abs(angle(f_1) - angle(f_2)) \leq \nu \wedge \theta_p^R > \theta_{left} \\ f_1 & otherwise \end{cases} \tag{5.26}$$

Here ν is the threshold within which a social selection strategy is invoked. If the best and 2nd best gaps are both competent, then the angular difference between their angles $abs(angle(f_1) - angle(f_2))$ is small. In such a case the social cues are followed, otherwise only the best gap is used. p is the nearest person and θ_p^R is the angle to the nearest person in the frame of reference of the robot. If the person is sufficiently to the right, then the avoidance is by right, otherwise, a left avoidance is preferred. The function *left*() selects the leftward of the two input angles as a social behaviour and *right*() selects the rightward of the two input angles.

Algorithm 5.2 - Gap Qualification Strategy

Input: laserscan $D = [d_i, \alpha_i]$
Output: qualified gaps: $gaps_Q$
Method:

1. $corners \leftarrow [0]$
2. For $i \in \{1,2, \ldots, len(D)\}$ do
   a. If $(abs(d_i - d_{i-1}) \leq \epsilon)$ then
      i. $corners.append(i-1)$
      ii. $corners.append(i)$
3. $corners.append(len(D))$
4. $gaps_Q = []$
5. For each gap f characterised by adjacent corner points (i, j) do
   a. $length(f) = d(f_{C1}, f_{C2})$
   b. $s_{distance} \leftarrow 0$
   c. $s_{angle} \leftarrow 0$
   d. For $k \in \{corners[i], corners[j]\}$ do
      i. $s_{distance} \leftarrow s_{distance} + d_k$
      ii. $s_{angle} \leftarrow s_{angle} + \alpha_k d_k$
   e. $angle(f) = \frac{s_{angle}}{s_{distance}}$
   f. $distance(f) = \frac{s_{distance}}{abs(corners[i] - corners[j] + 1)}$
   g. If $(length(f) \geq 2R \wedge distance(f) \geq \eta)$ then
      i. $gaps_Q.append(f)$
6. return $(gaps_Q)$

5.4.3 Overall Approach

The aim is to fuse goal seeking behaviour (asking the robot to go directly towards the goal) and gap seeking behaviour (going towards the chosen gap f_{SOC}). Nearer to the obstacles, the seeking gap behaviour dominates and away from the obstacles, the goal seeking dominates. The preferred angle is given by Equation (5.27)

$$\theta_{gap} = \frac{\frac{w}{\min(d_i)} angle(f_{soc}) + \theta_g}{\frac{w}{\min(d_i)} + 1} \tag{5.27}$$

Here $min(d_i)$ is the distance to the closest obstacle. w is an algorithm parameter such that $\frac{w}{\min(d_i)}$ is the relative weight of the gap seeking behaviour.

Algorithm 5.3 - Gap Selection Strategy

Input: qualified gaps: $gaps_Q$, front people ($P_F = \{p_f\}$), rear people ($P_R = \{p_r\}$), robot pose $q(x_r, y_r, \theta_r)$, goal(g)

Output: selected angle of motion θ_{des}

Method:

1. $f_1 = \arg\min_{f \in gaps_Q} angleDifference(angle(f) + \theta_r - \theta_g)$
2. $f_2 = \arg\min_{f \in gaps_Q, f \neq f_1} angleDifference(angle(f) + \theta_r - \theta_g)$
3. $p \leftarrow \arg\min_{p \in P_{front} \cup P_{rear}} d(q, p)$
4. $\theta_p^R \leftarrow$ orientation of person wrt the robot
5. If$(len(P_{front} \cup P_{rear}) \geq 1)$then
    a. If$abs(angle(f_1) - angle(f_2)) \leq v$ then
        i. If$(\theta_p^R \leq \theta_{left})$ then
            (A). $f \leftarrow left(f_1, f_2)$
        ii. Else
            (B). $f \leftarrow right(f_1, f_2)$
    b. Else
        i. $f_{soc} \leftarrow f_1$
6. Else
    a. $f_{soc} \leftarrow f_1$
7. If$(d(g, q) \leq distance(f_{soc}))$ then
    a. $\theta_{gap} \leftarrow \theta_g$
8. Else
    a. $\theta_{gap} \leftarrow \frac{\frac{w}{\min(d_i)} angle(f_{soc}) + \theta_g}{\frac{w}{\min(d_i)} + 1}$
9. return θ_{gap}

The algorithm for the gap selection is shown as Algorithm 5.2. The gap qualification strategy first extracts the corner points from the laser scan by taking the derivative of the adjacent points. The gaps are then obtained from the corner points based on a qualification criterion. The gap size must be larger than twice the size of the robot for it to qualify as an acceptable gap and the gap must be far enough. A deviation from the mathematical formulation is that the laser indices are used to denote the corners.

The gap selection is shown by Algorithm 5.3. The gap selection strategy processes the available gaps to select the ideal gap. The gap is chosen based on its direction with respect to the direction of the goal and the gap whose direction is closest to that of the goal is picked. The social cues are accounted for by giving a preference to the left-hand side when two gaps which are equally close to the goal are available in the presence of a human. A threshold is applied such as the potential field. If the gap is after the goal, it means a straight-line motion to the goal is possible and the robot selects the same.

5.5 Hybrid Force Model

The extended forces model accounts for the cooperation between the robot and the person through social cues; but the approach is reactive, meaning that the action taken to avoid the obstacle is after the obstacle is close enough to the robot. Reactive approaches tend to suffer from the problem of being stuck in a local minimum because they attempt to avoid collision only once the obstacle is sufficiently close enough to the robot. The extended gap method, while still a local planning algorithm, is more predictive compared to the force model. The robot attempts to move through obstacle free regions thus being more proactive by moving in obstacle free regions. The robot takes predictive actions to avoid the collisions. The quality of the trajectory generated however is dependent on the quality of the gaps detected which can sometimes be a tricky task. The lack of a reactive strategy in this method can sometimes lead to the robot generating a trajectory which moves too close to the human therefore encroaching their personal space and making it uncomfortable.

5.5.1 Fusion Design

The idea is to combine the reactive nature of the social force model with the proactive nature of the extended follow the gap method, such that the characteristics of both the methods are retained by the hybrid method. The qualification and selection strategy are integrated into the force model whose actions have both predictive and reactive properties. The predictive property is derived from the gap method where the robot tries to find the best obstacle free region and the reactive property is derived from the force model where the repulsion is very high when the obstacle is close and therefore preventing collisions.

The integration can occur in two ways. As a naïve way to integrate, the resultant angle of the force model can be combined with the gap angle using a single parameter to calculate the final heading angle which acts as the trade-off between the two methods. Consider F as the resultant total force, urging the robot to move in the direction $\theta_{APF}=atan2(F_y,F_x)$. Let the desired angle by the follow the gap method be $\theta_{gap}=angle(f_{soc})$. The desired resultant angle may thus be given by $(w_{APF}\ \theta_{APF}+ w_{gap}\ \theta_{gap})$, where w_{APF} and w_{gap} are the weights associated with the strategies. There are two problems with this approach. The first is that potential fields and follow the gap can have contradictory ways to avoid the obstacle, whose fusion can result in a collision. Say an obstacle somewhat ahead is seen. The first method suggests a leftward avoidance, while the second suggests a rightward avoidance, whose weighted addition results in a collision. Secondly, the relative weights are very hard to set.

The technique used is to integrate the selected gap directly into the force model as an attractive force, where the gap is modelled as a sub-goal to the main goal. The force from the main goal only comes into play when the distance from the destination to the robot is lesser than that of the selected gap from the robot. The gap selected by the selection strategy acts as a sub-goal for the robot and the attractive force is applied from the gap if it is closer to the robot compared to the goal. The attractive force is thus given by Equation (5.28).

$$F_{attr} = \begin{cases} k_{att} d(q,g) \begin{bmatrix} \cos(\theta_{gap}) \\ \sin(\theta_{gap}) \end{bmatrix} & \text{if } d(q,g) \le d_{max} \\ k_{att} d_{max} \begin{bmatrix} \cos(\theta_{gap}) \\ \sin(\theta_{gap}) \end{bmatrix} & \text{if } d(q,g) > d_{max} \end{cases} \tag{5.28}$$

Here θ_{gap} is the angle to gap as calculated by algorithm 5.3.

The social forces act in the same manner as previously in the force model. The social cues followed in the gap selection strategy works with the social forces to compound each other and thus result in safer trajectories. In this way, the gap strategy is integrated into the social force model in a very intuitive manner without any additional parameters and combines the proactive and reactive natures of both the methods.

Algorithm 5.4 - Hybrid Potential field algorithm

Input: laserscan ($D = [d_i, \alpha_i]$), selected gap angle (θ_{gap}), qualified gaps: $gaps_Q$, front people ($P_F = \{p_f\}$), rear people ($P_R = \{p_r\}$), robot pose $q(x_r, y_r, \theta_r)$, goal(g)
Output: angle of the motion
Method:

1. $F \leftarrow [0\ 0]^T$
2. For $< d_i, \alpha_i > \in D$ do
   a. $F \leftarrow F - k_{rep} \frac{1}{d_i^2} \begin{bmatrix} \cos(\theta_r + \alpha_i) \\ \sin(\theta_r + \alpha_i) \end{bmatrix}$
3. If$(d(q,g) \leq d_{max})$ then
   a. $F \leftarrow F + k_{att} d(q,g) \begin{bmatrix} \cos(\theta_{gap}) \\ \sin(\theta_{gap}) \end{bmatrix}$
4. else
   a. $F \leftarrow F + k_{att} d_{max} \begin{bmatrix} \cos(\theta_{gap}) \\ \sin(\theta_{gap}) \end{bmatrix}$
5. For $p \in P_{rear} \cup P_{front}$ do
   a. if$\theta_p^R \leq \theta_{left}$ then
      i. $F \leftarrow F - k_{soc} \frac{1}{d^2(q,p)} \begin{bmatrix} \cos\left(\theta_r + \theta_p^R - \frac{\pi}{2}\right) \\ \sin\left(\theta_r + \theta_p^R - \frac{\pi}{2}\right) \end{bmatrix}$
   b. else
      i. $F \leftarrow F - k_{soc} \frac{1}{d^2(q,p)} \begin{bmatrix} \cos\left(\theta_r + \theta_p^R + \frac{\pi}{2}\right) \\ \sin\left(\theta_r + \theta_p^R + \frac{\pi}{2}\right) \end{bmatrix}$
6. $\omega_{pref} = K_v \text{angleDifference}(\text{atan2}(F_y, F_x), \theta_r)$
7. if$\omega_{pref} \leq 0$ then
   a. $\omega \leftarrow \max(\omega_{pref}, -\omega_{max})$
8. else
   a. $\omega \leftarrow \min(\omega_{pref}, \omega_{max})$
9. $v = K_\omega \min(d_{front}(q), d(q,g))$
10. return (v, ω)

5.5.2 Overall Approach

The working of the algorithm is given in Algorithm 5.4. The social forces present in the extended force model remain unchanged and they come into play when people are detected moving towards the robot with the help of the cameras. The passing on the left convention is followed by default and can be easily changed by reversing the directions of the forces applied in case of a different convention. The selection strategy accounts for the social cues as well by factoring them in when choosing the best gap/sub-goal for the robot to move into. As the conventions followed by both the strategies are the same, their effects compound each other. For instance, when a person is walking towards the robot, the gap selection strategy enforces the convention and picks the gap on the left side of the person as its sub-goal. The social force arising from the person's presence is also directed to the left. The resultant of the forces ensures that the robot will pass on the left-hand side and maintain a safe distance from the person. In the potential field approach the robot moves forward until the pedestrian is at a sufficient distance, at which point it reacts to his or her presence. In the hybrid strategy, the robot starts moving towards a gap in a proactive manner and then makes a manoeuvre to maintain a safe distance from the person. In this way, the robot moves in a more human-like manner as well as maintains a larger distance from the pedestrians thereby respecting their private space.

5.6 Implementation

The system is implemented on the Pioneer LX robot. The Pioneer LX is equipped with a laser scanner to detect the obstacles around. Additionally, two Logitech webcams were mounted on top of the robot, one in front and one at the rear, in order to detect the presence of humans. This section discusses the different modules that complete the overall system architecture. The overall architecture representing the approach is shown in Figure 5.5.

Figure 5.5: Overall architecture of the hybrid model. The laser scan gives information about obstacles, used to calculate repulsion forces. The humans (detected by camera) are used to calculate social forces. The attraction forces are calculated by the goal. The local map produced by laser scan gives the gap, which produces a sub-goal for attraction. The forces determine direction of motion.

5.6.1 Face Detection and Tracking

The facenet model (Schroff et al., 2015), which is the state-of-the-art model for face detection and recognition, is used for this purpose. Here, a correlation tracker is used to track the detected face for a certain number of frames, before performing detection once again because it has been observed that the Logitech web camera does not perform very well for the face detection when it is in motion.

5.6.2 Face Distance and Angle Computation

Once the bounding box of the face is obtained, the area of the bounding box is used to calculate the distance and orientation of the person from the robot. The area of the bounding box is observed to be inversely proportional to the distance of the person from the robot.

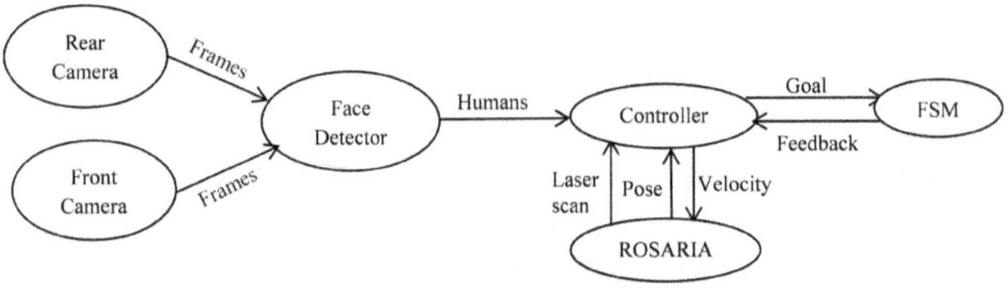

Figure 5.6: ROS Graph for the problem. The vertices represent software modules. The edges represent the information interchanged. The ROSARIA is the manufacturer supplied software suite for interacting with the robot.

Sample data was collected by placing subjects at different distances from the robot and then measuring the area of the bounding box detected. The angle of the person was observed as well. The camera had a 60-degree field of view and pixel location of the bounding box was used to calculate the angle. A linear regression model was trained on the collected data to predict the distance of a person, given the size of the bounding box of his/her face and its orientation. The model gave acceptable results upon further testing and was hence used to detect the positions of the humans.

5.6.3 Sequencing

The problem of reaching a goal directly may not be possible in a reactive methodology because of the presence of complex obstacles on the way. The reactive algorithms can often get stuck in complex obstacle scenarios. Therefore, the problem of motion to goal is decomposed by breaking a deliberative path to goal into several sub-goals. The reactive control algorithms are only provided with the sub-goals. As soon as the robot reaches a sub-goal, within a tolerance level, the sequencer provides the next sub-goal to the robot. In this way, on progressing forward, the robot keeps getting the next sub-goals, until it reaches near enough to the goal, when it is given the goal.

The overall system was implemented using the Robot Operating System (ROS) (Quigley et al., 2009). The different modules executed as different nodes of ROS, which communicated via ROS messages. A representative ROS graph is given by Figure 5.6.

5.7 Results

The different approaches were tested across multiple scenarios in both the simulations and on a real robot and the distance maintained from the humans was used as a metric to compare the performances of the approaches.

5.7.1 Evaluation Metric

A penalty metric as a key performance indicator is designed factoring in the socially acceptable distance. Penalties are given to the robot throughout the course of the trial when it violates the distance constraint. The metric is given by Equation (5.29).

$$\text{penalty}(t) = \begin{cases} \frac{0.1}{d(q,p)} & \text{if } d(q,p) \leq d_{soc} \\ 0 & \text{otherwise} \end{cases} \tag{5.29}$$

Here d_{soc} is the maximum social distance as observed by humans, which was taken as 0.5m and used for minimising the penalty for violation of social distance between the robot and humans. The metric of 0.5m was derived as follows. Experiments were carried out by observing humans in natural environments. A large room was used for the study, so that the environmental bias is extremely small. People were made to avoid each other. The people were tracked using a Quanergy 3D lidar mounted on a tripod. The 3D image at every time step was processed to remove background, remove outliers, perform orthographic projection, and cluster to separate the individuals. The detected individuals were then subjected to filtering by using a Kalman filter. The separation of people at different times, specifically during avoiding each other were noted. The experiments revealed an avoidance clearance of 0.2401m±0.0382m. In the experiments, the distance between the centres of the robot and person is used as the clearance. Thus, the desired clearance is (approximately) 0.2401m (clearance)+0.25m (radius of the robot) = 0.4901m. Note that the separation is measured by a lidar which touches the obstacle (person) and therefore the radius of the person is not excluded. An additional factor of 0.01m is left since the human may not be as comfortable with the robot with another human.

The four approaches used for testing are given as follows. The APF method is the basic social force model without any social cues and tends to suffer from an oscillatory behaviour. The Follow The Gap Social Cues (FTG-SC) is the extended follow the gap method with the social cues. The Social Potential Field Social Cues (SPF-SC) is the extended force model with the social cues. The Hybrid Social Cues (Hybrid-SC) method is the is the approach integrating the gap selection strategy into the force model.

5.7.2 Evaluation of the social force

The motivation was that the people prefer to avoid each other on the left-side (unless the human is not already too much on the left). To ascertain the same rule, social experiments were carried out by the author. Cameras were fitted in one of the shopping centres of the city where the people flock in large numbers. The videos were studied to observe the motion of every human. Stress was given to the specific situations where the humans come head-on towards each other. 50 such scenarios were manually mined. In 38 scenarios, the passing by was on the left, while in 12 the passing by was on the right. The right side is primarily selected when there is no space in the left and humans' goal is also on right side. The passing on the left therefore has a confidence given by Equation (5.30).

$$\text{Confidence} = \frac{|\text{left}_{\text{pass}}|}{|\text{left}_{\text{pass}}|+|\text{right}_{\text{pass}}|} \tag{5.30}$$

$$\text{Confidence} = \frac{38}{38+12} = 0.76$$

Some snapshots of the passing by are given in Figure 5.7. That said, in the Indian scenario the humans usually follow the left side rule in navigation, which is different in other countries and the conventions need to be interchanged. Let us consider the case where the humans do not keep to the left-hand side. There is always a very brief period where the humans are nearly head-to-head. The human choosing to go on the right is typically very confident and takes a very quick drift. This breaks the uncertainty region in the algorithm about the side of avoidance and the robot shall counter by taking the opposite direction. Here the robot moves much slower than the people in real life and therefore the robot barely moves in the time when the person shows their strong intent to pass via the opposite side.

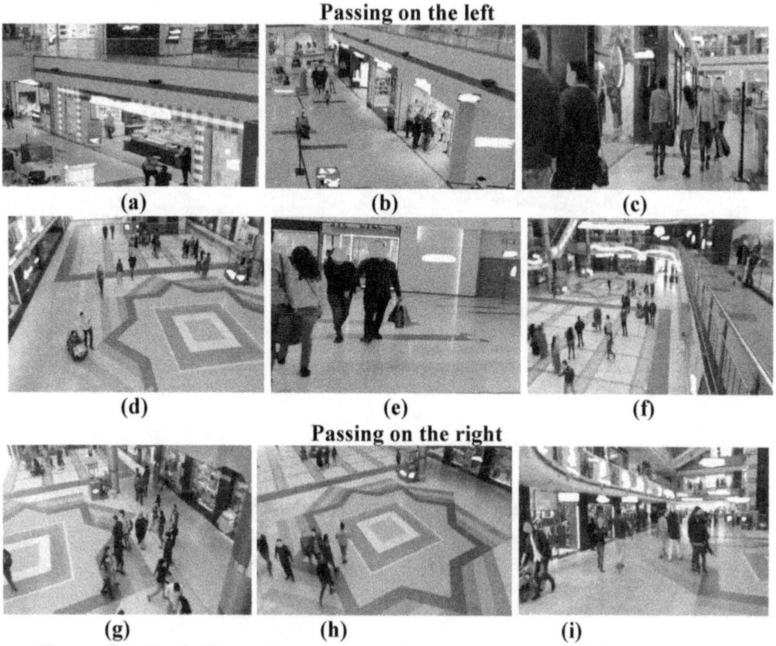

Figure 5.7: Social Experiments to study passing on the left hypothesis

One of the contributions of this chapter was to add a social force in the working of the potential field algorithm. In order to socially study the effect of this force, this thesis caried out a social experiment. Two subjects were asked to walk towards each other naturally and their motion was observed using a 3D lidar. The trajectories are shown in Figure 5.8 that show ability of the algorithm to imitate human trajectory available in the human trajectory database. The simulation was then carried out using the potential field approach with and without the social force and the results are shown in Figure 5.8. Clearly, the social force was better able to imitate the real observed trajectory. Without the social force, the two robots drift apart only when they are very close to each other, while with the social force the robots have in-anticipation given enough space to each other. This proves the effectiveness of the social force.

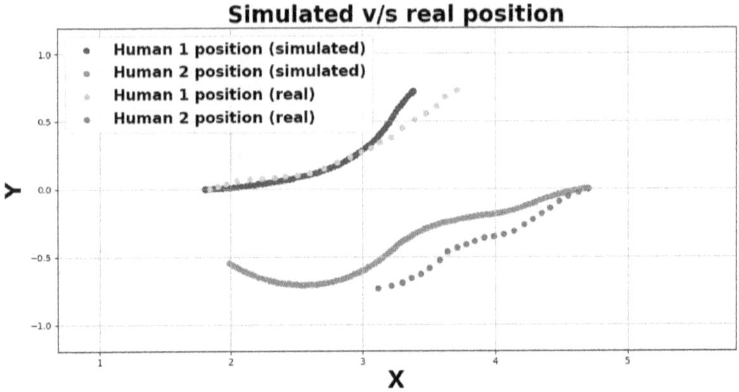

(a) With social force

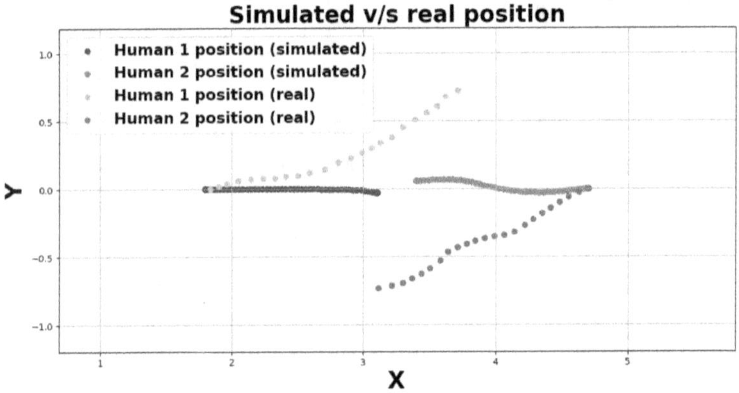

(b) Without social force

Figure 5.8: Effect of the social force on the robot

5.7.3 Simulations

All the methods have been implemented and tested using the ROS framework on a Linux operating system with configuration 12GB RAM, CORE I5 processor. MobileSim is the simulator which was used for developing and testing the algorithms, before being tested on the real robot. The map used by the simulator was from the Centre of Intelligent Robotics at IIIT Allahabad. The map was made by the Pioneer LX robot using the onboard lidar sensor and a SLAM algorithm.

The methods have been tested extensively in various scenarios on simulations. In the simulations, there are multiple instances of the robot that move according to the same algorithm and therefore each acts as a moving human for the other robots to apply the navigational strategies to move in a safe manner. The Pioneer LX robot is used in the simulations. The scenarios involve similar situations to that of the real-life navigation of the robots passing each other as well as overtaking scenarios when a robot is overtaken by another. The approaches were tested in the simulations in four different scenarios three of which consisted of at least three dynamic obstacles in the robot's vicinity while the fourth was a static scenario not consisting of dynamic obstacles. In the first scenario, each of the robots have their goals at around 5 metres in front of them. As they move towards their goals, they encounter dynamic obstacles and social cues are used to avoid them. The scenario is shown in Figure 5.9(a). In the second scenario, the leftmost robot is deliberately placed a little below the middle robot to make the motion a bit more complicated. The rightmost robot after having passed the first robot on the left has to go towards its right instead of left as there is no space on the left. The scenario is shown in Figure 5.9(b). In the third scenario, the robot has multiple sub-goals which are to first move 5 metres straight ahead which is the first sub-goal and then turn right and move 4 metres which is the second sub-goal. Along the way, dynamic obstacles are encountered, and social cues are used to avoid them. The scenario is shown in Figure 5.9(c). In the fourth (static) scenario, the robot has to travel around the entire laboratory. The entire task is broken down into multiple sub goals and the robot moves its current position to its next sub goal. The FSM acts as the global planner and provides the next goal to the robot once the current goal is reached. The entire trajectory is planned such that the robot takes a tour of the entire laboratory. The scenario is shown in Figure 5.9 (d). The minimum clearance maintained by the robot throughout the trial, the time taken to complete the task as well as the penalty incurred by the robot were recorded and presented in Tables 5.1-5.3.

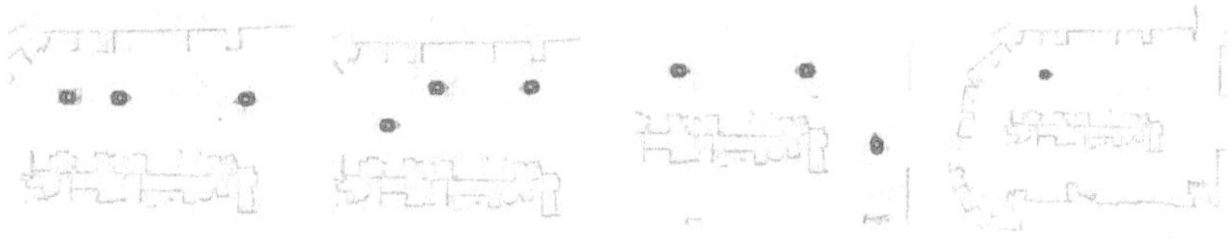

(a) Scenario 1 **(b) Scenario 2** **(c) Scenario 3** **(d) Scenario 4**

Figure 5.9: Scenarios used for testing

The performance of the algorithm is judged by the key performance indicators that are clearance maintained by the robot, time and the penalty obtained. Adherence of the social norms is not explicitly discussed since in all the scenarios passing on the left and overtaking on the right social norms are observed. The higher the clearance and lower the penalty, the better is the algorithm. From the above observations, it can be inferred that the algorithms accounting for the social cues outperform their non-social counterparts in generating safe trajectories in a pedestrian rich environment. The baseline approach results in the robot moving too close to the human being in many cases and is therefore unsafe. The clearance increases as social cues are introduced into the different methods making the navigation safer. In most of the scenarios, the novel hybrid algorithm Hybrid-SC, which is an integration of the geometric approach into the extended social potential fields, has the best performance as it maintains a higher clearance than the other methods on a consistent basis. The novel hybrid algorithm maintains the highest clearance compared to the other methods in all the scenarios. It also has the smallest penalties obtained in all but one scenario and hence generates safer trajectories in most cases. The SPF-SC method, which is the social force model extended to account for social forces, has a smaller penalty in one scenario compared to the Hybrid model. The hybrid model being an extension of the force model integrated with the gap selection strategy therefore has proactive characteristics unlike the force field models which are purely reactive. Hence, the novel hybrid method combines the best features of the force model and geometric approaches to outperform the previous model-based methods by generating safer trajectories.

Table 5.1: Clearances maintained in meters

Clearance (m)	**Scenario 1**	**Scenario 2**	**Scenario 3**	**Static Scenario**
APF	0.31	0.28	0.32	0.92
FTG-SC	0.35	0.41	0.29	0.84
SPF-SC	0.45	0.43	0.41	0.92
Hybrid-SC	**0.56**	**0.48**	**0.44**	**0.95**

5.7.4 Hardware Experiments

The algorithm was also implemented on the Pioneer LX robot operating at the Centre of Intelligent Robotics at the Institute. The robot is operated using ROS. The robot consists of a lidar sensor that is used to detect the obstacles around as well as for localisation. A camera is placed on top of the robot to capture the real time video in order to detect the presence of humans. The face detection algorithm (Zhang et al., 2016) was used to detect the presence of humans in the robot surroundings. As the robot is operating in a known environment, its position at any given time is obtained through localisation using a manufacturer supplied library in ROS. The localisation, face detection, motion planning and other modules interact as ROS nodes. The scenarios in which the approaches were tested involves the presence of multiple humans. In these experiments, the robot starts from a starting position and is expected to move sequentially towards all sub-goal positions in the environment. As humans are present in this environment, the robot navigates towards its sub-goals by maintaining enough clearance and respecting the social conventions. The scenarios usually involve humans walking towards the robot or across the robot causing the social forces from the humans to act on the robot and hence allowing the robot to safely navigate by avoiding collisions. Students working in the laboratory were asked to walk past/towards the robot multiple times.

Table 5.2: Penalty obtained

Penalty	**Scenario 1**	**Scenario 2**	**Scenario 3**	**Static Scenario**
APF	12.52	17.8	23.6	0
FTG-SC	7.5	10.8	25.4	0.2
SPF-SC	5.72	**6.4**	20.7	0
Hybrid-SC	**0**	7.0	**20.1**	**0**

Table 5.3: Time taken in seconds

Time Taken (s)	**Scenario 1**	**Scenario 2**	**Scenario 3**	**Static Scenario**
APF	15.10	16.75	31.16	78.3
FTG-SC	15.02	14.85	29.66	84.3
SPF-SC	15.29	15.73	30.86	78.3
Hybrid-SC	15.13	15.43	30.56	83.5

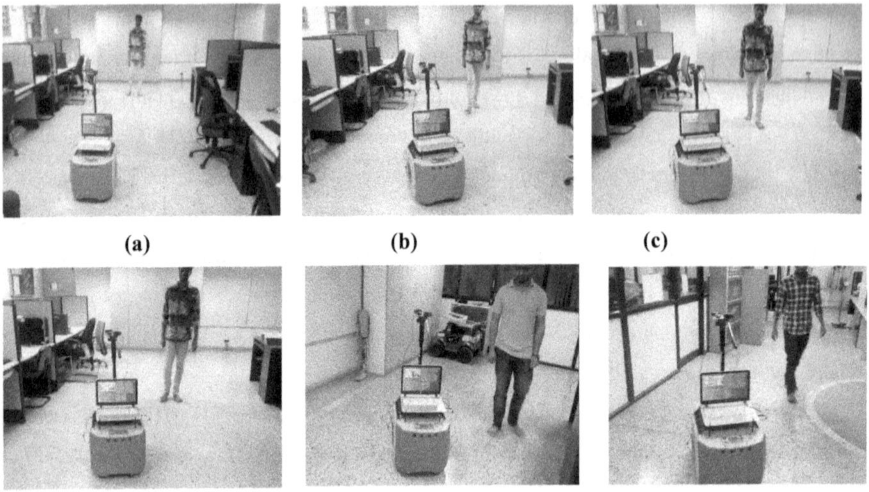

(a) (b) (c)

(d) (e) (f)

Figure 5.10: Execution of the algorithm on the Pioneer LX robot

A few snapshots of the run are shown in Figure 5.10. The results show the robot first avoids the first human from the correct side, for which the robot had an ample of time and the intention of the person was also clear. However, soon after avoiding the person, the robot had to turn around the corner where it suddenly sees the second person. Ideally while turning the robot would have centrally placed itself, however, encountering the person at the corner, the robot suddenly drifts leftward to give a larger social distance to the person. Thereafter, after another turn, the robot meets a person who has taken a sharp motion towards the robot's right and the robot acknowledges this by taking a left. The clearances and angles of the robot during the run are plotted as Figure 5.11.

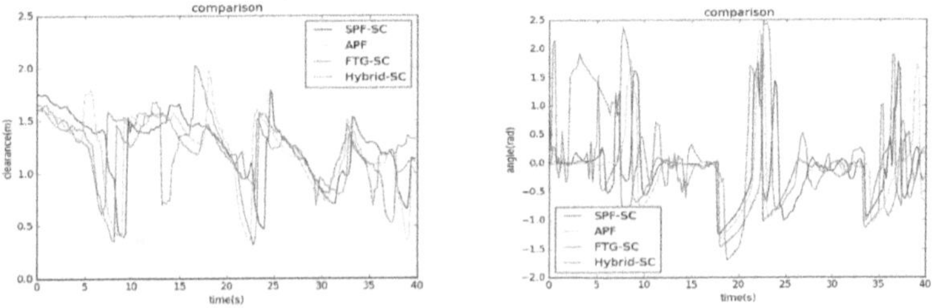

(a) (b)

Figure 5.11: Metrics on the run with the real robot

The comparison with the different approaches is shown in Table 5.4. Just as observed in the simulations, the Hybrid-SC method outperformed its counterparts even more significantly in the experiments with the real robot. The minimum clearance maintained by the robot from the obstacles was 0.56 meters and more than 10 centimetres higher than the next best method. The method did not receive any penalty as the minimum distance constraint was never violated. The novel hybrid method has shown to be superior to the previous approaches in both the simulations as well as the real-world experiments on the robot. The hybrid model being an extension of the force model integrated with the gap selection strategy therefore has proactive characteristics unlike the force field models which are purely reactive. Hence, the novel hybrid method combines the best features of the force model and geometric approaches to outperform the previous model-based methods by generating safer trajectories.

Table 5.4: Real World Scenario

Pioneer LX	Clearance (m)	Time Taken (s)	Penalty
APF	0.34	52.99	59.1
FTG-SC	0.29	51.3	56.7
SPF-SC	0.41	41.3	50
Hybrid-SC	**0.56**	**41.09**	**0**

5.7.5 Additional Simulations

The real-world settings take a long time and the requirement of a reasonable number of human actors. Further testing is therefore undertaken using simulations. In simulations multiple instances of robots running under the proposed algorithms is used to analyse the working of the algorithm. The simulations are used to generate many diverse situations that are otherwise seen in the everyday navigation within the laboratory. At the same time, the corner cases of the algorithms including robots ahead of each other are used in the simulations to ensure that the algorithms act appropriately. The simulations enable the placing of several robots and running the algorithm to observe the behaviour. The simulations are a good way to ensure that the robots avoid each other with enough clearance and do not block each other or end up in a deadlock, which is possible for robots.

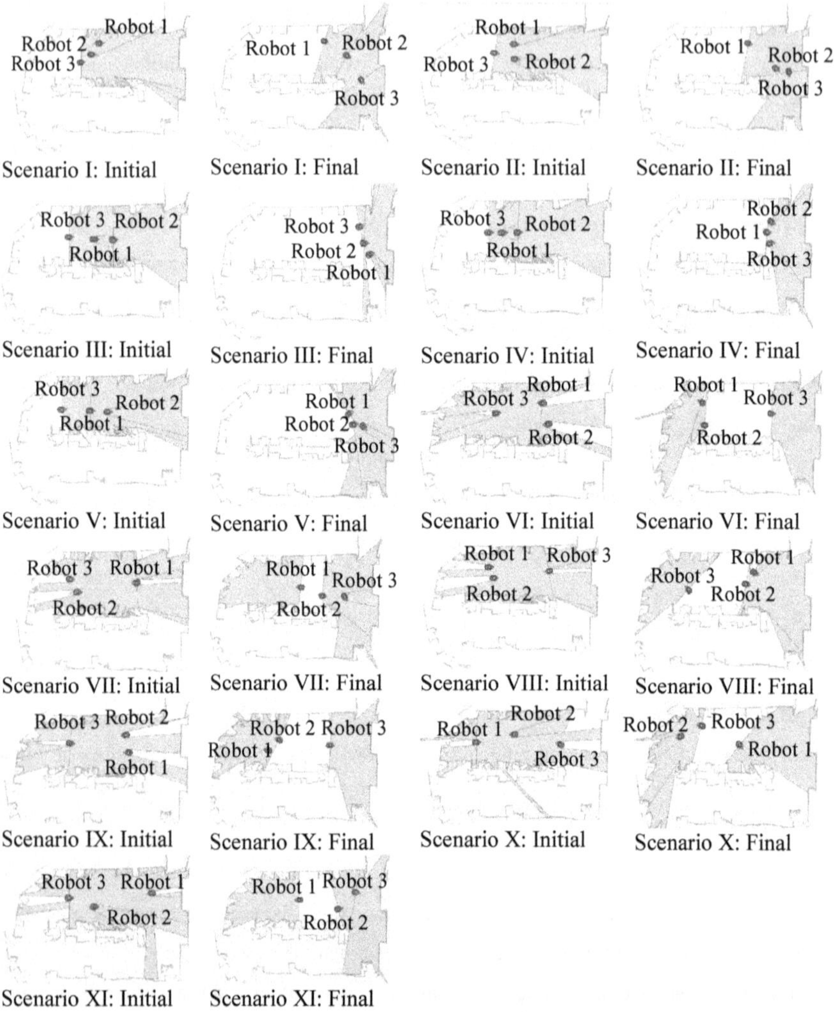

Figure 5.12: Simulation Results

The testing carried out done on 11 scenarios with 3 robots each. The first 4 scenarios are for overtaking and showing cooperation into overtaking behaviours. The last 7 scenarios test avoidance of robots in different settings. Overtaking may also happen due to speed differences. The variations of speeds and initial inter-robot separation is carried out in all scenarios. Some of the scenarios are under an adversarial setting where there is a very small distance for the robot to pass by. In scenario I, robot 3 is aggressive and overtakes both the robots. To make matters difficult, robot 2 also later attempts to overtake robot 1. In scenario II, robot 3 is to overtake robot 1 and robot 2 which are together ahead. Robot 3 makes the robots drift to the left to enable overtaking. In scenario III, while robot 1 is made to overtake robot 2, the other robots can merely go towards their goal. In scenario IV, the robots are in a line and the challenge is to laterally displace while moving longitudinally towards their goal. In scenario V, while robot 3 overtakes, robot 1 is given a goal opposite to the general convention and robot 2 must hence adjust. Scenario VI features robot 1 and robot 2 with a very high lateral displacement with robot 3 approaching from the other side.

Table 5.5: Simulation Results on different Scenarios

Scenario	Clearance (m)			Penalty			Time (s)		
	Robot 1	Robot 2	Robot 3	Robot 1	Robot 2	Robot 3	Robot 1	Robot 2	Robot 3
I	0.59	0.60	0.45	0.0	0.0	10.78	27.75	23.57	24.17
II	0.55	0.83	0.42	0.0	0.0	13.89	22.06	22.67	26.90
III	0.33	1.02	0.27	7.07	0.0	10.932	28.41	16.62	14.21
IV	0.5	1.38	0.48	14.8	0.0	17.46	24.78	16.62	30.83
V	0.49	0.51	0.23	11.13	0.0	9.00	22.37	17.22	12.39
VI	0.5	0.55	0.5	10.0	0.0	9.2	16.32	16.92	15.11
VII	0.39	0.33	0.34	14.42	15.27	18.50	22.07	23.27	25.08
VIII	0.5	0.32	0.28	12.6	10.69	10.37	22.36	17.53	13.6
IX	0.42	0.22	0.23	10.90	19.14	61.11	21.76	21.15	54.11
X	0.5	0.5	0.5	8.2	11.0	31.0	22.06	16.93	54.13
XI	0.26	0.29	0.31	17.25	6.88	1.81	22.07	17.23	10.27

The social conventions are simultaneously followed for robot 2 and not followed for robot 3 because of the goal. Robot 3 detects the same and choses the correct side for avoidance, while also carefully balancing the forces. Scenario VII tightens the conditions of Scenario VI by choosing goals with a smaller lateral displacement. Scenario VIII makes robot 2 simultaneously avoid both robot 3 and robot 1 while it is head-on to both of them. In scenario IX, robot 3 must essentially avoid robot 2, while robot 1 makes the lateral separations bound for the two robots. Scenario X makes robot 1 avoid robot 2 from the non-social side, and immediately places robot 3 in front which must also be similarly avoided. Finally, scenario XI makes robot 3 avoid robot 2 easily, which however makes it get head-on with robot 1, which must be quickly avoided. The metrics of the scenarios are shown in Table 5.5. The initial and the final positions are shown in Figure 5.12.

5.7.6 Analysis

The two basic behaviours presented are robot avoidance when the two robots face each other and cooperative robot overtaking. The behaviours are affected by two factors, the speeds of the robots and the inter-robot distances. Let us therefore analyse the algorithm functioning over these factors. The first behaviour is when two robots face each other. Experiments are carried out by varying the speeds and the results are presented in Figure 5.13. As the speed increase, the robots get a much shorter time to orient themselves, which reduces the clearance and increases the penalty. The increasing speed, obviously, reduces the travel time. Due to symmetry, the metrics of the two robots are nearly the same, and can be seen as overlapping plots. There is a little deviation due to the stochasticity of the experiments. Similarly, the inter-robot distance is changed and the results are shown in Figure 5.14. With a larger inter-robot distance, the robots get more time and hence get larger clearance and smaller penalty. There is no major difference in travel time though.

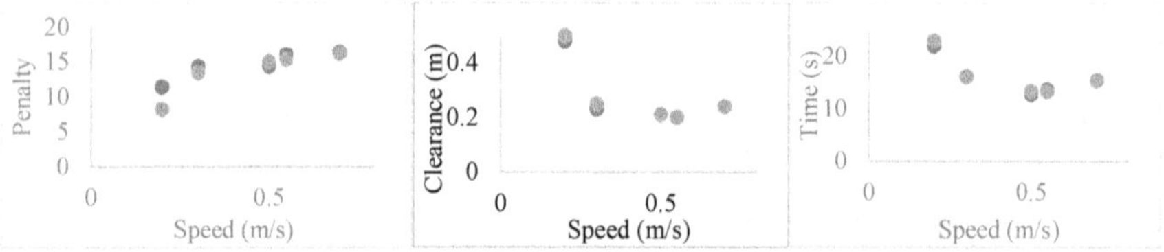

Figure 5.13: Analysis of robot avoidance behaviour by changing speeds

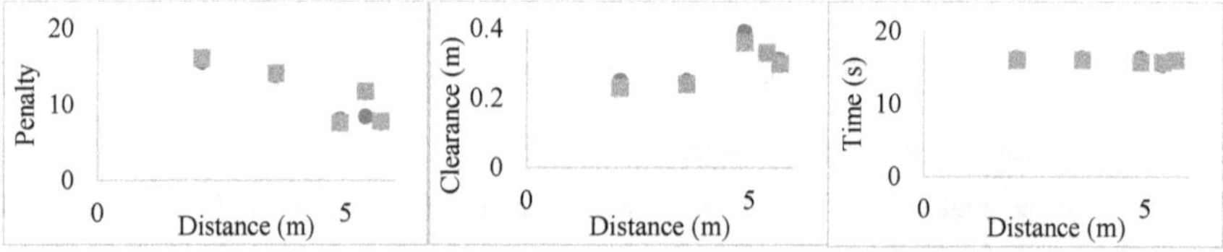

Figure 5.14: Analysis of robot avoidance behaviour by changing inter-robot distance

The other behaviour is overtaking. First experiments are made by varying the speed. Higher speed gives a lesser time for adjusting towards a comfortable overtaking distance, which reduces the clearance and increases the penalty for the overtaking robot. The time shows a decreasing trend. The results are shown in Figure 5.15. The inter-robot distance is also varied as shown in Figure 5.16. If the robots have a large distance, there is more space to adjust to a good overtaking position, which increases the clearance and reduces the penalty. The travel time is reasonably stable. The values for the overtaken robot are due to the limited field of view prior to overtaking and hence the overtaken robot was not aware of the overtaking robot behind at the time of closest proximity.

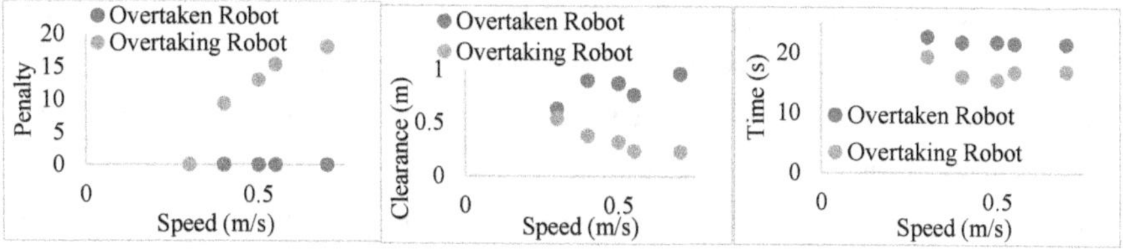

Figure 5.15: Analysis of overtaking behaviour by changing speeds

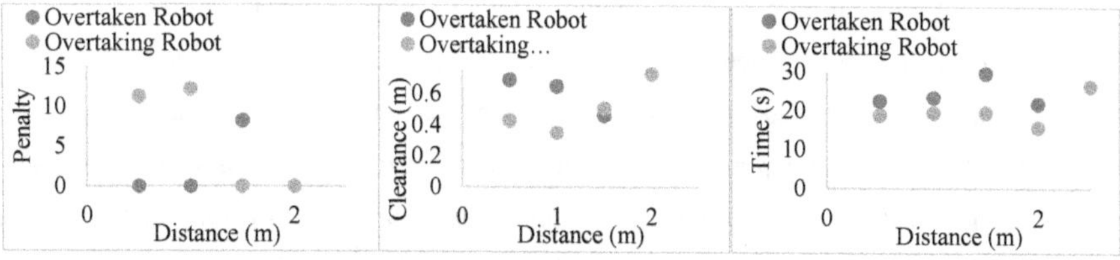

Figure 5.16: Analysis of overtaking behaviour by changing inter-robot distance

5.8 Conclusion

This chapter presents a social robot motion planning algorithm for service robots that operate amidst humans. In general, artificial potential field is used for motion planning of the mobile robot based on attraction and repulsion forces. The idea of social cues is implemented on top of the artificial potential field algorithm which is inspired from the Indian traffic conditions and pedestrian walking patterns. The modified approach is named as the social potential field with social cues or SPF-SC. Here if a person is available at the front of the robot, then the robot will go from the left side, although the robot can go from the right side if there is no space at the left side. This algorithm gives the linear and angular velocity to the robot. In this algorithm the robot may suddenly change its direction which is also observed in traffic scenarios. Consider a person driving a vehicle and suddenly another vehicle or obstacle comes from the front side, then the person will immediately change the direction of the vehicle which is considered very risky. If the person already knew from which side the other vehicle is coming, then the change in direction is anticipatory. Therefore, a follow the gap behaviour is also implemented for the motion planning of the mobile robot, which is made socially aware by making the robot prefer the left side for avoiding an oncoming person. The approach is named as follow the gap with social cues or FTW-SC.

The two approaches are hybridised as an approach called as Hybrid-SC. In the hybrid approach, a repulsion force is applied for each obstacle. Alongside, the best gap is selected using the social norms for achieving the goal. An attraction force is applied in the direction of the gap. Additionally, if a person is available at the front of the robot, a repulsion force is applied in the left direction of the robot, else in the right direction. The total force gives the linear and angular velocity of the robot. In the artificial potential field algorithm, the robot avoids the obstacles only when they are very close to the robot, and it also suddenly takes a turn due to repulsion which creates a risk. In the follow the gap method the robot initially determines all the possible gaps where the robot can enter and achieve its goal and uses a proactive mechanism to avoid obstacles while seeking for the selected gap. However, if an obstacle comes very close at the front of the robot, then the follow the gap method may make the robot travel still closer to the obstacle in pursuit of the selected gap, while the artificial potential field approach readily makes the robot drift away from the obstacle, making potential field a better algorithm. The contradictory pros and cons of both approaches (FTG-SC and SPF-SC) are merged in the Hybrid-SC, which solves both the problem that arise in FTG-SC and SPF-SC.

Each method is tested on one static scenario and three different dynamic scenarios on a simulator followed by the real robot in live settings. The average clearance maintained by the robot is 0.92 meters, 0.84 meters, 0.92 meters, and 0.95 meters using the APF, FTG-SC, SPF-SC, and Hybrid-SC algorithms respectively for the static scenario, showing the improvement of the hybrid approach over the two counterparts. The clearance improvement by the proposed method is 3.16% (APF), 11.58% (FTG-SC), and 3.16% (SPF-SC) in the static scenario. Clearance is better as compared to all the other approaches in the static scenario because the moving robot initially detected the obstacle and nothing else is moving. The additional social repulsion force is not applied as the obstacle is static. Attraction towards the socially selected gap makes the robot proactively make a larger clearance. The benefits are amplified in the dynamic scenarios that pose a greater uncertainty due to the presence of the other agents. The clearances recorded are 0.30

meters, 0.35 meters, 0.43 meters, and 0.49 meters from APF, FTG-SC, SPF-SC, and Hybrid-SC respectively for the dynamic scenario, showing the hybrid approach better by a larger margin. The improvement is 38.77% (APF), 28.58% (FTG-SC), and 12.25% (SPF-SC) in the dynamic scenario. The betterment gets even further amplified over the live settings on a real robot, where the clearances are 0.34 meters, 0.29 meters, 0.41 meters, and 0.56 meters from APF, FTG-SC, SPF-SC, and Hybrid-SC respectively. The proposed approach now has a significantly larger margin of improvement with a betterment of 39.29% (APF), 48.22% (FTG-SC), and 26.79% (SPF-SC) in the real-live settings. In the dynamic scenario and real-life settings, the clearance is improved by Hybrid-SC because APF takes a decision when the robot is very close to the obstacle while in the Hybrid-SC, the gap is computed at each iteration and the robot takes a decision for changing its direction as soon as the obstacles are detected.

A penalty metric is designed factoring in the socially acceptable distance. Penalties are given to the robot through the course of the trial when it violates the distance constraint. The average penalty obtained by the robot is 0, 0.2, 0, and 0 from APF, FTG-SC, SPF-SC, and Hybrid-SC respectively for the static scenario. The penalty obtained is approximately 0 for all the approaches in the static scenario. The obstacle is static and therefore all the methods keep enough distance to avoid a penalty. The penalties come into play in the dynamic scenario. The penalties are 17.97, 14.56, 10.94, and 9.03 from APF, FTG-SC, SPF-SC, and Hybrid-SC respectively for the dynamic scenario. The penalty reduction by the proposed algorithm is 49.74% (APF), 37.98% (FTG-SC), and 17.45% (SPF-SC) in the dynamic scenario. The real live settings further amplify the penalty metric. The penalties received by the methods are 59.1, 56.7, 50, and 0 for APF, FTG-SC, SPF-SC, and Hybrid-SC respectively. In the dynamic scenario and real-live settings, the penalty obtained by the proposed hybrid approach is zero or very less as compared to the other approaches. In the APF, the robot avoids an obstacle only when it is very close to it, so the safety distance is violated and hence the robot obtains a higher penalty. Although, the proposed Hybrid-SC approach obtained a zero penalty in the real-life settings, it obtained a non-zero penalty for the dynamic setting because the selected gap did not have enough space with the obstacle already into it, while the corridor was narrow, and the robot was circumventing a turn.

The average time taken by the robot was also measured for all the approaches. For the static scenario, the time taken by the robot is 78.3 seconds, 84.3 seconds, 78.3 seconds, and 83.5 seconds from APF, FTG-SC, SPF-SC, and Hybrid-SC respectively. For the dynamic settings the time taken by the robot reduces to 21.00 seconds, 19.84 seconds, 20.62 seconds, and 20.37 seconds from APF, FTG-SC, SPF-SC, and Hybrid-SC respectively. For the real robot experiments the time taken by the robot is 52.99 seconds, 51.3 seconds, 41.3 seconds, and 41.09 seconds from APF, FTG-SC, SPF-SC, and Hybrid-SC respectively.

There are a few limitations that are important to mention. The approach is reactive in nature and does not guarantee the robot to be trap-free when operating in complex maps. The approach will require the assistance of a deliberative planner. The real-world settings used human face detection to enable the robot to apply the social cues. The false positives and false negatives in the detection can adversely affect the algorithm. The current decision making does not account for the speeds of the different agents that can be an important factor to account for in the future.

REFERENCES

Abbeel, P., & Ng, A. Y. (2004, July). Apprenticeship learning via inverse reinforcement learning. In *Proceedings of the twenty-first international conference on Machine learning* (p. 1).

Aghaei, M., Dimiccoli, M., & Radeva, P. (2016). Multi-face tracking by extended bag-of-tracklets in egocentric photo-streams. *Computer Vision and Image Understanding, 149*, 146-156.

Ahonen, T., Hadid, A., & Pietikäinen, M. (2004, May). Face recognition with local binary patterns. In *European conference on computer vision* (pp. 469-481). Springer, Berlin, Heidelberg.

Alač, M, Movellan, J, Tanaka, F (2011) When a robot is social: Spatial arrangements and multimodal semiotic engagement in the practice of social robotics. *Social Studies of Science* 41(6): 893-926.

Arulampalam, M. S., Maskell, S., Gordon, N., & Clapp, T. (2002). A tutorial on particle filters for online nonlinear/non-Gaussian Bayesian tracking. *IEEE Transactions on signal processing*, 50(2), 174-188.

Bay, H., Ess, A., Tuytelaars, T., & Van Gool, L. (2008). Speeded-up robust features (SURF). *Computer vision and image understanding*, 110(3), 346-359.

Bandini, S., Federici, M. L., Manzoni, S., & Vizzari, G. (2005, October). Towards a methodology for situated cellular agent based crowd simulations. In *International Workshop on Engineering Societies in the Agents World* (pp. 203-220). Springer, Berlin, Heidelberg.

Bencherif, A., & Chouireb, F. (2019) A recurrent TSK interval type-2 fuzzy neural networks control with online structure and parameter learning for mobile robot trajectory tracking. *Applied Intelligence*, 49(11), 3881-3893.

Bennewitz, M., Burgard, W., Cielniak, G., & Thrun, S. (2005). Learning motion patterns of people for compliant robot motion. *The International Journal of Robotics Research*, 24(1), 31-48.

Bera, A., & Manocha, D. (2014) Realtime multilevel crowd tracking using reciprocal velocity obstacles. In *2014 22nd International Conference on Pattern Recognition*, pp 4164-4169.

Bera, A., Randhavane, T., Prinja, R., & Manocha, D. (2017, September) Sociosense: Robot navigation amongst pedestrians with social and psychological constraints. In *2017 IEEE/RSJ International Conference on Intelligent Robots and Systems (IROS)* (pp. 7018-7025). IEEE.

Bevilacqua, P., Frego, M., Fontanelli, D., & Palopoli, L. (2018) Reactive planning for assistive robots. *IEEE Robotics and Automation Letters*, 3(2), 1276-1283.

Bimbo, A.D., & Dini, F. (2011). Particle filter-based visual tracking with a first order dynamic model and uncertainty adaptation. *Computer Vision and Image Understanding*, 115(6), 771-786.

Borenstein, J., & Koren, Y. (1991). The vector field histogram-fast obstacle avoidance for mobile robots. *IEEE transactions on robotics and automation*, 7(3), 278-288.

Breazeal, C, Dautenhahn, K, Kanda, T (2016) Social Robotics. In: Siciliano B., Khatib O. (eds) *Springer Handbook of Robotics*. Springer Handbooks. Springer, Cham.

Brown, S., & Waslander, S. L. (2016, October) The constriction decomposition method for coverage path planning. In *2016 IEEE/RSJ International Conference on Intelligent Robots and Systems (IROS)* (pp. 3233-3238). IEEE.

Burgard, W., Cremers, A. B., Fox, D., Hähnel, D., Lakemeyer, G., Schulz, D., Steiner W., & Thrun, S. (1998, July). The interactive museum tour-guide robot. In *Aaai/iaai* (pp. 11-18).

Burgard, W., Cremers, A. B., Fox, D., Hähnel, D., Lakemeyer, G., Schulz, D., Steiner W., & Thrun, S. (1999). Experiences with an interactive museum tour-guide robot. *Artificial intelligence*, 114(1-2), 3-55.

Burgoon, JK, Jones, SB (1976). Toward a theory of personal space expectations and their violations. *Human Communication Research*, 2(2):131–146, 1976.

Castillo, J. C., Fernández-Caballero, A., Serrano-Cuerda, J., López, M. T., & Martínez-Rodrigo, A. (2017) Smart environment architecture for robust people detection by infrared and visible video fusion. *Journal of Ambient Intelligence and Humanized Computing*, 8(2), 223-237.

Chen, G., Cui, G., Kong, L., Guo, S., & Cao, L. (2019) Particle-filter-based human target tracking in image domain for through-wall imaging radar. *The Journal of Engineering*, 2019(20), 6549-6553.

Chen, Q., & Luh, J. Y. S. (1994, May). Coordination and control of a group of small mobile robots. In *Proceedings of the 1994 IEEE International Conference on Robotics and Automation* (pp. 2315-2320). IEEE.

Chen, X., Wang, X., & Xuan, J. (2018). Tracking multiple moving objects using unscented Kalman filtering techniques. *arXiv preprint arXiv:1802.01235.*

Chen, Y. F., Liu, M., Everett, M., & How, J. P. (2017, May). Decentralized non-communicating multiagent collision avoidance with deep reinforcement learning. In *2017 IEEE international conference on robotics and automation (ICRA)* (pp. 285-292). IEEE.

Chen, Y. F., Everett, M., Liu, M., & How, J. P. (2017, September). Socially aware motion planning with deep reinforcement learning. In *2017 IEEE/RSJ International Conference on Intelligent Robots and Systems (IROS)* (pp. 1343-1350). IEEE.

Chi, W., Wang, J., & Meng, M. Q. H. (2017) A gait recognition method for human following in service robots. *IEEE Transactions on Systems, Man, and Cybernetics: Systems*, *48*(9), 1429-1440.

Choi, C., & Christensen, H. I. (2011, May) Robust 3D visual tracking using particle filtering on the SE (3) group. In *2011 IEEE International Conference on Robotics and Automation* (pp. 4384-4390). IEEE.

Choi, C., & Christensen, H. I. (2012) Robust 3D visual tracking using particle filtering on the special Euclidean group: A combined approach of keypoint and edge features. *The International Journal of Robotics Research,* 31(4), 498-519.

Curtis, S., Best, A., & Manocha, D. (2016). Menge: A modular framework for simulating crowd movement. *Collective Dynamics,* 1, 1-40.

Dahroug, B., Tamadazte, B., & Andreff, N. (2017, May) Visual servoing controller for time-invariant 3d path following with remote centre of motion constraint. In *2017 IEEE International Conference on Robotics and Automation (ICRA)* (pp. 3612-3618). IEEE.

Dautenhahn, K., Woods, S., Kaouri, C., Walters, M. L., Koay, K. L., & Werry, I. (2005, August). What is a robot companion-friend, assistant or butler?. In *2005 IEEE/RSJ international conference on intelligent robots and systems* (pp. 1192-1197). IEEE.

Demir, M., & Sezer, V. (2017, July). Improved Follow the Gap Method for obstacle avoidance. In *2017 IEEE International Conference on Advanced Intelligent Mechatronics (AIM)* (pp. 1435-1440). IEEE.

Deo, N., Rangesh, A., & Trivedi, M. M. (2018) How would surround vehicles move? a unified framework for maneuver classification and motion prediction. *IEEE Transactions on Intelligent Vehicles,* 3(2), 129-140.

Duan, Y., Lu, J., Feng, J., & Zhou, J. (2018). Topology preserving structural matching for automatic partial face recognition. *IEEE Transactions on Information Forensics and Security*, 13(7), 1823-1837.

Everett, M., Chen, Y. F., & How, J. P. (2018, October). Motion planning among dynamic, decision-making agents with deep reinforcement learning. In *2018 IEEE/RSJ International Conference on Intelligent Robots and Systems (IROS)* (pp. 3052-3059). IEEE.

Feil-Seifer, D., & Mataric, M. J. (2005, June). Defining socially assistive robotics. In *9th International Conference on Rehabilitation Robotics, 2005. ICORR 2005.* (pp. 465-468). IEEE.

Ferraguti, F., Landi, C. T., Secchi, C., Fantuzzi, C., Nolli, M., & Pesamosca, M. (2017) Walk-through programming for industrial applications. *Procedia Manufacturing, 11*, 31-38.

Ferrer, G, Garrell, A, Sanfeliu, A (2013) Robot companion: A social force based approach with human awareness-navigation in crowded environments. In: *IEEE/RSJ International Conference on Intelligent Robots and Systems,* pp. 1688-1694.

Ferrer, G., Garrell, A., & Sanfeliu, A. (2013, September). Social-aware robot navigation in urban environments. In *2013 European Conference on Mobile Robots* (pp. 331-336). IEEE.

Fiorini, L., Mancioppi, G., Semeraro, F., Fujita, H., & Cavallo, F. (2020) Unsupervised emotional state classification through physiological parameters for social robotics applications. *Knowledge-Based Systems*, *190*, 105217.

Fourie, J., Mills, S., & Green, R. (2010). Harmony filter: a robust visual tracking system using the improved harmony search algorithm. *Image and Vision Computing*, *28*(12), 1702-1716.

Franchi, A., Freda, L., Oriolo, G., & Vendittelli, M. (2007, April). A randomized strategy for cooperative robot exploration. In *Proceedings 2007 IEEE International Conference on Robotics and Automation* (pp. 768-774). IEEE.

Fulgenzi, C., Spalanzani, A., & Laugier, C. (2007, April). Dynamic obstacle avoidance in uncertain environment combining PVOs and occupancy grid. In *Proceedings 2007 IEEE International Conference on Robotics and Automation* (pp. 1610-1616). IEEE.

Garrell, A., Sanfeliu, A., & Moreno-Noguer, F. (2009, October). Discrete time motion model for guiding people in urban areas using multiple robots. In *Intelligent Robots and Systems, 2009. IROS 2009. IEEE/RSJ International Conference on* (pp. 486-491). IEEE.

Gao, M. L., Li, L. L., Sun, X. M., & Luo, D. S. (2014). Face tracking based on differential harmony search. *IET Computer Vision*, 9(1), 98-109.

Gao, R., Zhao, M., Ye, T., Ye, F., Wang, Y., & Luo, G. (2017). Smartphone-based real time vehicle tracking in indoor parking structures. *IEEE Transactions on Mobile Computing*, 16(7), 2023-2036.

Gopinath, D, Jain, S, Argall, BD (2017) Human-in-the-Loop Optimization of Shared Autonomy in Assistive Robotics. *IEEE Robotics and Automation Letters* 2(1):247-254.

Görür, O. C., & Erkmen, A. M. (2014, August). Elastic networks in reshaping human intentions by proactive social robot moves. In *The 23rd IEEE international symposium on robot and human interactive communication* (pp. 1012-1017). IEEE.

Guo, S., Xu, H., Thalmann, N. M., & Yao, J. (2017). Customization and fabrication of the appearance for humanoid robot. *The Visual Computer*, 33(1), 63-74.

Haritaoglu, I., Harwood, D., & Davis, L. S. (1999, September). Hydra: Multiple people detection and tracking using silhouettes. In *Proceedings 10th International Conference on Image Analysis and Processing* (pp. 280-285). IEEE.

Haidegger, T., Kovács, L., Precup, R. E., Preitl, S., Benyó, B., & Benyó, Z. (2011) Cascade control for telerobotic systems serving space medicine. *IFAC Proceedings Volumes*, 44(1), 3759-3764.

Hayashi, K., Sakamoto, D., Kanda, T., Shiomi, M., Koizumi, S., Ishiguro, H., Ogasawara T., & Hagita, N. (2007, March). Humanoid robots as a passive-social medium-a field experiment at a train station. In *Human-Robot Interaction (HRI), 2007 2nd ACM/IEEE International Conference on* (pp. 137-144). IEEE.

He, K., Zhang, X., Ren, S., & Sun, J. (2015) Delving deep into rectifiers: Surpassing human-level performance on imagenet classification. In *Proceedings of the IEEE international conference on computer vision*, pp. 1026-1034.

Helble, H., & Cameron, S. (2007, April). 3-d path planning and target trajectory prediction for the oxford aerial tracking system. In *Proceedings 2007 IEEE International Conference on Robotics and Automation* (pp. 1042-1048). IEEE.

Helbing, D., & Molnar, P. (1995). Social force model for pedestrian dynamics. *Physical review E*, 51(5), 4282.

Henriques, J. F., Caseiro, R., Martins, P., & Batista, J. (2014) High-speed tracking with kernelized correlation filters. *IEEE transactions on pattern analysis and machine intelligence*, 37(3), 583-596.

Hornby, G. S., Takamura, S., Yokono, J., Hanagata, O., Yamamoto, T., & Fujita, M. (2000). Evolving robust gaits with AIBO. In *Robotics and Automation, 2000. Proceedings. ICRA'00. IEEE International Conference on* (Vol. 3, pp. 3040-3045). IEEE.

Ishiguro, H., Ono, T., Imai, M., Maeda, T., Kanda, T., & Nakatsu, R. (2001). Robovie: an interactive humanoid robot. *Industrial robot: An international journal*, 28(6), 498-504.

Jang, S., Elmqvist, N., & Ramani, K. (2015) Motionflow: Visual abstraction and aggregation of sequential patterns in human motion tracking data. *IEEE transactions on visualization and computer graphics*, 22(1), 21-30.

Jeddisaravi, K., Alitappeh, R. J., Pimenta, L. C., & Guimaraes, F. G. (2016) Multi-objective approach for robot motion planning in search tasks. *Applied Intelligence*, 45(2), 305-321.

Jensfelt, P., Austin, D. J., Wijk, O., & Andersson, M. (2000) Feature based condensation for mobile robot localization. In *Proceedings 2000 ICRA. Millennium Conference. IEEE International Conference on Robotics and Automation,* Vol. 3, pp. 2531-2537.

Kala, R. (2018). On repelling robotic trajectories: coordination in navigation of multiple mobile robots. *Intelligent Service Robotics*, 11(1), 79-95.

Kala, R. (2018). Routing-based navigation of dense mobile robots. *Intelligent Service Robotics,* 11(1), 25-39.

Kala, R, Warwick, K (2013) Motion Planning of Autonomous Vehicles in a Non-Autonomous Vehicle Environment without Speed Lanes. *Engineering Applications of Artificial Intelligence,* 26(5-6): 1588–1601.

Kaplan, F. (2000, October). Talking AIBO: First experimentation of verbal interactions with an autonomous four-legged robot. In *Learning to behave: interacting agents CELE-TWENTE Workshop on Language Technology* (pp. 57-63).

Khaksar, W., Hong, T. S., Khaksar, M., & Motlagh, O. (2014) A fuzzy-tabu real time controller for sampling-based motion planning in unknown environment. *Applied intelligence*, 41(3), 870-886.

Khare, A., Motwani, R., Akash, S., Patil, J., & Kala, R. (2018, July). Learning the goal seeking behaviour for mobile robots. In *2018 3rd Asia-Pacific Conference on Intelligent Robot Systems (ACIRS)* (pp. 56-60). IEEE.

Khatib, O. (1985, March). Real-time obstacle avoidance for manipulators and mobile robots. In *Proceedings. 1985 IEEE International Conference on Robotics and Automation* (Vol. 2, pp. 500-505). IEEE.

Khatib, O. (1986). Real-time obstacle avoidance for manipulators and mobile robots. *In Autonomous robot vehicles* (pp. 396-404). Springer, New York, NY.

Kim, B., & Pineau, J. (2016) Socially adaptive path planning in human environments using inverse reinforcement learning. *International Journal of Social Robotics*, 8(1), 51-66.

Knepper, R. A., & Rus, D. (2012, September). Pedestrian-inspired sampling-based multi-robot collision avoidance. In *2012 IEEE RO-MAN: The 21st IEEE International Symposium on Robot and Human Interactive Communication* (pp. 94-100). IEEE.

Knobloch, A., Vahrenkamp, N., Wächter, M., & Asfour, T. (2018) Distance-Aware Dynamically Weighted Roadmaps for Motion Planning in Unknown Environments. *IEEE Robotics and Automation Letters*, *3*(3), 2016-2023.

Kratz, L., & Nishino, K. (2011). Tracking pedestrians using local spatio-temporal motion patterns in extremely crowded scenes. *IEEE transactions on pattern analysis and machine intelligence*, 34(5), 987-1002.

Kretzschmar, H., Spies, M., Sprunk, C., & Burgard, W. (2016) Socially compliant mobile robot navigation via inverse reinforcement learning. *The International Journal of Robotics Research*, 35(11), 1289-1307.

Krizhevsky, A., Sutskever, I., & Hinton, G. E. (2012) Imagenet classification with deep convolutional neural networks. In *Advances in neural information processing systems,* pp. 1097-1105.

Kruse, T., Pandey, A. K., Alami, R., & Kirsch, A. (2013). Human-aware robot navigation: A survey. *Robotics and Autonomous Systems*, 61(12), 1726-1743.

Kumar, N., & Sukavanam, N. (2020) An improved CNN framework for detecting and tracking human body in unconstraint environment. *Knowledge-Based Systems*, 193, 105198.

Kumar, U., Banerjee, A., & Kala, R. (2020) Collision avoiding decentralized sorting of robotic swarm. *Applied Intelligence*, 50(4), 1316-1326.

Landi, C. T., Ferraguti, F., Secchi, C., & Fantuzzi, C. (2016, October) Tool compensation in walk-through programming for admittance-controlled robots. In *IECON 2016-42nd Annual Conference of the IEEE Industrial Electronics Society* (pp. 5335-5340). IEEE.

Large, F., Vasquez, D., Fraichard, T., & Laugier, C. (2004, June). Avoiding cars and pedestrians using velocity obstacles and motion prediction. In *IEEE Intelligent Vehicles Symposium, 2004* (pp. 375-379). IEEE.

Lenser, S., & Veloso, M. (2000) Sensor resetting localization for poorly modelled mobile robots. In Proceedings 2000 ICRA. *Millennium Conference. IEEE International Conference on Robotics and Automation,* Vol. 2, pp. 1225-1232.

Li, G. H., Chang, C. F., & Fu, L. C. (2008) Navigation of a wheeled mobile robot in indoor environment by potential field based-fuzzy logic method. In *2008 IEEE Workshop on Advanced robotics and Its Social Impacts* pp 1-6.

Li, H., Lin, Z., Shen, X., Brandt, J., & Hua, G. (2015) A convolutional neural network cascade for face detection. *In Proceedings of the IEEE conference on computer vision and pattern recognition,* pp. 5325-5334.

Li, S. Z., & Zhang, Z. (2004). Floatboost learning and statistical face detection. *IEEE Transactions on pattern analysis and machine intelligence*, 26(9), 1112-1123.

Li, Y., Wang, Z., Yang, X., Wang, M., Poiana, S. I., Chaudhry, E., & Zhang, J. (2019). Efficient convolutional hierarchical autoencoder for human motion prediction. *The Visual Computer*, 35(6), 1143-1156.

Liu, Z., Huang, J., Han, J., Bu, S., & Lv, J. (2016) Human motion tracking by multiple RGBD cameras. *IEEE Transactions on Circuits and Systems for Video Technology*, 27(9), 2014-2027.

Long, P., Liu, W., & Pan, J. (2017). Deep-learned collision avoidance policy for distributed multiagent navigation. *IEEE Robotics and Automation Letters*, *2*(2), 656-663.

Luber, M., Spinello, L., Silva, J., & Arras, K. O. (2012, October). Socially-aware robot navigation: A learning approach. In *2012 IEEE/RSJ International Conference on Intelligent Robots and Systems* (pp. 902-907). IEEE.

Luber, M., Stork, J. A., Tipaldi, G. D., & Arras, K. O. (2010, May). People tracking with human motion predictions from social forces. In *2010 IEEE International Conference on Robotics and Automation* (pp. 464-469). IEEE.

Luo, J., & Qin, S. (2018) A fast algorithm of simultaneous localization and mapping for mobile robot based on ball particle filter. *IEEE Access,* 6, 20412-20429.

Marchetti, F., Becattini, F., Seidenari, L., & Del Bimbo, A. (2020) Multiple Trajectory Prediction of Moving Agents with Memory Augmented Networks. *IEEE Transactions on Pattern Analysis and Machine Intelligence.*

Mata, M., Armingol, J. M., de la Escalera, A., & Salichs, M. A. (2001, May). A visual landmark recognition system for topological navigation of mobile robots. In *Proceedings 2001 ICRA. IEEE International Conference on Robotics and Automation (Cat. No. 01CH37164)* (Vol. 2, pp. 1124-1129). IEEE.

Lou, J., Tan, T., Hu, W., Yang, H., & Maybank, S. J. (2005). 3-D model-based vehicle tracking. *IEEE Transactions on image processing*, 14(10), 1561-1569.

Martinez-Garcia, E. A., Akihisa, O., Yuta, S. (2005, June). Crowding and guiding groups of humans by teams of mobile robots. In *Advanced Robotics and its Social Impacts, 2005. IEEE Workshop on* (pp. 91-96). IEEE.

Meir, R., & Rätsch, G. (2003). An introduction to boosting and leveraging. In *Advanced lectures on machine learning* (pp. 118-183). Springer, Berlin, Heidelberg.

Min, W., Fan, M., Guo, X., & Han, Q. (2017). A new approach to track multiple vehicles with the combination of robust detection and two classifiers. *IEEE Transactions on Intelligent Transportation Systems*, 19(1), 174-186.

Murphy, R (2000) The Hybrid Deliberative/Reactive Paradigm. In *Introduction to AI Robotics, Second Edition*, MIT Press, Cambridge, MA.

Ng, A. Y., & Russell, S. J. (2000, June). Algorithms for inverse reinforcement learning. In *Icml* (Vol. 1, p. 2).

Orozco-Rosas, U., Picos, K., & Montiel, O. (2019) Hybrid Path Planning Algorithm Based on Membrane Pseudo-Bacterial Potential Field for Autonomous Mobile Robots. *IEEE Access*, 7, 156787-156803.

Özdemir, A., & Sezer, V. (2018). Follow the gap with dynamic window approach. *International Journal of Semantic Computing,* 12(01), 43-57.

Paliwal, S. S., & Kala, R. (2018) Maximum clearance rapid motion planning algorithm. Robotica, 36(6), 882-903.

Pandey, AK, Alami, R (2010) A framework towards a socially aware mobile robot motion in human-centered dynamic environment. In *2010 IEEE/RSJ International Conference on Intelligent Robots and Systems,* pp. 5855-5860.

Park, J. S., Park, C., & Manocha, D. (2019) I-Planner: Intention-aware motion planning using learning-based human motion prediction. *The International Journal of Robotics Research*, 38(1), 23-39.

Parkhi, O.M., Vedaldi A., Zisserman A. (2015). *Deep Face Recognition*. BMVC, 1(3), 1-12.

Pereira, P., Cunha, R., Cabecinhas, D., Silvestre, C., & Oliveira, P. (2019) A 3-D Trailer Approach to Leader-Following Formation Control. *IEEE Transactions on Control Systems Technology*.

Pozna, C., Troester, F., Precup, R. E., Tar, J. K., & Preitl, S. (2009) On the design of an obstacle avoiding trajectory: Method and simulation. *Mathematics and Computers in Simulation*, 79(7), 2211-2226.

Pradhan, N., Burg, T., & Birchfield, S. (2011, June). Robot crowd navigation using predictive position fields in the potential function framework. In *Proceedings of the 2011 American control conference* (pp. 4628-4633). IEEE.

Pransky, J. (2004). Social adjustments to a robotic future. *Wolf and Mallett*, 137-59.

Quigley, M, Gerkey, B, Conley, K, Faust, J, Foote, F, Leibs, J., Berger, E, Wheeler, R, Ng, A. (2009) ROS: an open-source Robot Operating System. *In ICRA workshop on open source software*, vol. 3, No. 3.2, p. 5.

Ratsamee, P., Mae, Y., Ohara, K., Takubo, T., & Arai, T. (2013). Human–robot collision avoidance using a modified social force model with body pose and face orientation. *International Journal of Humanoid Robotics*, 10(01), 1350008.

Reif, J. H., & Wang, H. (1999). Social potential fields: A distributed behavioral control for autonomous robots. *Robotics and Autonomous Systems*, 27(3), 171-194.

Repiso, E., Garrell, A., & Sanfeliu, A. (2019). Adaptive side-by-side social robot navigation to approach and interact with people. *International Journal of Social Robotics*, 1-22.

Reynolds, C. W. (1987, August). Flocks, herds and schools: A distributed behavioral model. In *Proceedings of the 14th annual conference on Computer graphics and interactive techniques* (pp. 25-34).

Rodriguez, M., Ali, S., & Kanade, T. (2009, September). Tracking in unstructured crowded scenes. In *2009 IEEE 12th International Conference on Computer Vision* (pp. 1389-1396). IEEE.

Roozegar, M., Mahjoob, M. J., Esfandyari, M. J., & Panahi, M. S. (2016) XCS-based reinforcement learning algorithm for motion planning of a spherical mobile robot. *Applied Intelligence*, 45(3), 736-746.

Schirner, G, Erdogmus, D, Chowdhury, K, Padir, T (2013) The Future of Human-in-the-Loop Cyber-Physical Systems. *Computer* 46(1): 36-45.

Schreier, M., Willert, V., & Adamy, J. (2014) Bayesian, maneuver-based, long-term trajectory prediction and criticality assessment for driver assistance systems. In *17th International IEEE Conference on Intelligent Transportation Systems*, pp 334-341.

Schroff, F., Kalenichenko, D., & Philbin, J. (2015). Facenet: A unified embedding for face recognition and clustering. In *Proceedings of the IEEE conference on computer vision and pattern recognition* (pp. 815-823).

Sezer, V., & Gokasan, M. (2012). A novel obstacle avoidance algorithm:"Follow the Gap Method". *Robotics and Autonomous Systems*, 60(9), 1123-1134.

Shan, M., Worrall, S., Masson, F., & Nebot, E. (2014). Using delayed observations for long-term vehicle tracking in large environments. *IEEE Transactions on Intelligent Transportation Systems*, 15(3), 967-981.

Shi, X., Wu, J., Ling, X., Zheng, Q. L., Pan, X. Q., & Zhao, Z. Q. (2016). Real-time face recognition method based on the threshold determination of the positive face sequence. In *Proceedings of the 22nd International Conference on Industrial Engineering and Engineering Management 2015* (pp. 125-136). Atlantis Press, Paris.

Shiomi, M., Kanda, T., Glas, D. F., Satake, S., Ishiguro, H., & Hagita, N. (2009, October). Field trial of networked social robots in a shopping mall. In *Intelligent Robots and Systems, 2009. IROS 2009. IEEE/RSJ International Conference on* (pp. 2846-2853). IEEE.

Shiomi, M., Kanda, T., Ishiguro, H., & Hagita, N. (2010, March). A larger audience, please!: encouraging people to listen to a guide robot. In *Proceedings of the 5th ACM/IEEE international conference on Human-robot interaction* (pp. 31-38). IEEE Press.

Shiomi, M., Kanda, T., Koizumi, S., Ishiguro, H., & Hagita, N. (2008). Group attention control for communication robots. *International Journal of Humanoid Robotics*, 5(04), 587-608

Shiomi, M., Zanlungo, F., Hayashi, K., & Kanda, T. (2014). Towards a socially acceptable collision avoidance for a mobile robot navigating among pedestrians using a pedestrian model. *International Journal of Social Robotics*, 6(3), 443-455.

Sisbot, E. A., Alami, R., Siméon, T., Dautenhahn, K., Walters, M., & Woods, S. (2005, December). Navigation in the presence of humans. In *Humanoid Robots, 2005 5th IEEE-RAS International Conference on* (pp. 181-188). IEEE.

Sisbot, EA, Marin-Urias, LF, Alami, R, Simeon, T (2007) A Human Aware Mobile Robot Motion Planner. *IEEE Transactions on Robotics* 23(5): 874-883.

Sun, L., Yan, Z., Mellado, S. M., Hanheide, M., & Duckett, T. (2018, May) 3DOF pedestrian trajectory prediction learned from long-term autonomous mobile robot deployment data. In *2018 IEEE International Conference on Robotics and Automation (ICRA)* (pp. 1-7). IEEE.

Sun, Y., Wang, X., & Tang, X. (2014). Deep learning face representation from predicting 10,000 classes. In *Proceedings of the IEEE conference on computer vision and pattern recognition* (pp. 1891-1898).

Sung, Y., & Chung, W. (2015) Hierarchical sample-based joint probabilistic data association filter for following human legs using a mobile robot in a cluttered environment. *IEEE Transactions on Human-Machine Systems*, 46(3), 340-349.

Svenstrup, M., Bak, T., & Andersen, H. J. (2010, October). Trajectory planning for robots in dynamic human environments. In *2010 IEEE/RSJ International Conference on Intelligent Robots and Systems* (pp. 4293-4298). IEEE.

Sviestins, E., Mitsunaga, N., Kanda, T., Ishiguro, H., & Hagita, N. (2007, March). Speed adaptation for a robot walking with a human. In *Proceedings of the ACM/IEEE international conference on Human-robot interaction* (pp. 349-356).

Takahashi, M., Suzuki, T., Cinquegrani, F., Sorbello, R., & Pagello, E. (2009, December). A mobile robot for transport applications in hospital domain with safe human detection algorithm. In *2009 IEEE International Conference on Robotics and Biomimetics (ROBIO)* (pp. 1543-1548). IEEE.

Taigman, Y., Yang, M., Ranzato, M. A., & Wolf, L. (2014). Deepface: Closing the gap to human-level performance in face verification. In *Proceedings of the IEEE conference on computer vision and pattern recognition* (pp. 1701-1708).

Tang, H., Sun, W., Yu, H., Lin, A., Xue, M., & Song, Y. (2019) A novel hybrid algorithm based on PSO and FOA for target searching in unknown environments. *Applied Intelligence*, 49(7), 2603-2622.

Thabet, E., Khalid, F., Sulaiman, P. S., & Yaakob, R. (2018) Fast marching method and modified features fusion in enhanced dynamic hand gesture segmentation and detection method under complicated background. *Journal of Ambient Intelligence and Humanized Computing*, 9(3), 755-769.

Thompson, S., Horiuchi, T., & Kagami, S. (2009, February). A probabilistic model of human motion and navigation intent for mobile robot path planning. In *2009 4th International Conference on Autonomous Robots and Agents* (pp. 663-668). IEEE.

Tian, S., Yin, X. C., Su, Y., & Hao, H. W. (2017) A unified framework for tracking based text detection and recognition from web videos. *IEEE transactions on pattern analysis and machine intelligence*, 40(3), 542-554.

Toit, N. E.D., & Burdick, J. W. (2010, May). Robotic motion planning in dynamic, cluttered, uncertain environments. In *2010 IEEE International Conference on Robotics and Automation* (pp. 966-973). IEEE.

Trautman, P., Ma, J., Murray, R. M., & Krause, A. (2015) Robot navigation in dense human crowds: Statistical models and experimental studies of human–robot cooperation. *The International Journal of Robotics Research*, 34(3), 335-356.

Truong, X. T., & Ngo, T. D. (2016). Dynamic social zone based mobile robot navigation for human comfortable safety in social environments. *International Journal of Social Robotics*, 8(5), 663-684.

Truong, X. T., & Ngo, T. D. (2017). Toward socially aware robot navigation in dynamic and crowded environments: A proactive social motion model. *IEEE Transactions on Automation Science and Engineering*, 14(4), 1743-1760.

Tsuda, N, Harimoto, S, Saitoh, T, Konishi, R (2009) Mobile robot with following and returning mode. In: *The 18th IEEE International Symposium on Robot and Human Interactive Communication*, Toyama, pp. 933-938.

Ulicny, B., & Thalmann, D. (2002, December). Towards Interactive Real-Time Crowd Behavior Simulation. In *Computer Graphics Forum* (Vol. 21, No. 4, pp. 767-775). Oxford, UK: Blackwell Publishing, Inc.

Ulrich, I., & Borenstein, J. (1998, May). VFH+: Reliable obstacle avoidance for fast mobile robots. In *Proceedings. 1998 IEEE international conference on robotics and automation (Cat. No. 98CH36146)* (Vol. 2, pp. 1572-1577). IEEE.

Ulrich, I., & Borenstein, J. (2000, April). VFH/sup*: Local obstacle avoidance with look-ahead verification. In *Proceedings 2000 ICRA. Millennium Conference. IEEE International Conference on Robotics and Automation. Symposia Proceedings (Cat. No. 00CH37065)* (Vol. 3, pp. 2505-2511). IEEE.

Van Den Berg, J., Guy, S. J., Lin, M., & Manocha, D. (2011). Reciprocal n-body collision avoidance. In *Robotics research* (pp. 3-19). Springer, Berlin, Heidelberg.

Van den Berg, J., Lin, M., & Manocha, D. (2008, May). Reciprocal velocity obstacles for real-time multi-agent navigation. In *2008 IEEE International Conference on Robotics and Automation* (pp. 1928-1935). IEEE.

Vaughan, R., Sumpter, N., Henderson, J., Frost, A., & Cameron, S. (2000). Experiments in automatic flock control. *Robotics and autonomous systems*, 31(1-2), 109-117.

Viola, P., & Jones, M. (2001, December). Rapid object detection using a boosted cascade of simple features. In *Proceedings of the 2001 IEEE computer society conference on computer vision and pattern recognition. CVPR 2001* (Vol. 1, pp. I-I). IEEE.

Wang, A., & Steinfeld, A. (2020) Group Split and Merge Prediction With 3D Convolutional Networks. *IEEE Robotics and Automation Letters*, 5(2), 1923-1930.

Wang, M., Liu, Y., Su, D., Liao, Y., Shi, L., Xu, J., & Miro, J. V. (2018) Accurate and real-time 3-D tracking for the following robots by fusing vision and ultrasonar information. *IEEE/ASME Transactions on Mechatronics*, 23(3), 997-1006.

Wang, P. K. (1991). Navigation strategies for multiple autonomous mobile robots moving in formation. *Journal of Robotic Systems*, 8(2), 177-195.

Weiss, A, Bader, M, Vincze, M, Hasenhütl, G, Moritsch, S (2014) Designing a service robot for public space: an action and experiences-approach. In: *Proceedings of the 2014 ACM/IEEE international conference on Human-robot interaction.* 2014.

Wen, F., Yuan, K., Zou, W., Chai, X., & Zheng, R. (2009, August). Visual navigation of an indoor mobile robot based on a novel artificial landmark system. In *2009 International Conference on Mechatronics and Automation* (pp. 3775-3780). IEEE.

Wen, Y., Zhang, K., Li, Z., & Qiao, Y. (2016, October) A discriminative feature learning approach for deep face recognition. In *European conference on computer vision* (pp. 499-515). Springer, Cham.

Weng, S. K., Kuo, C. M., & Tu, S. K. (2006). Video object tracking using adaptive Kalman filter. *Journal of Visual Communication and Image Representation*, 17(6), 1190-1208.

Wengefeld, T., Müller, S., Lewandowski, B., & Gross, H. M. (2019, October) A Multi Modal People Tracker for Real Time Human Robot Interaction. In *2019 28th IEEE International Conference on Robot and Human Interactive Communication (RO-MAN)* (pp. 1-8). IEEE.

Xie, G., Gao, H., Qian, L., Huang, B., Li, K., & Wang, J. (2017) Vehicle trajectory prediction by integrating physics-and maneuver-based approaches using interactive multiple models. *IEEE Transactions on Industrial Electronics*, 65(7), 5999-6008.

Xu, L. Q., Landabaso, J. L., & Lei, B. (2004). Segmentation and tracking of multiple moving objects for intelligent video analysis. *BT technology Journal*, 22(3), 140-150.

Yu, T., Zhao, J., Huang, Y., Li, Y., & Liu, Y. (2019) Towards Robust and Accurate Single-View Fast Human Motion Capture. *IEEE Access*, 7, 85548-85559.

Yu, Y., Mora, K. A. F., & Odobez, J. M. (2018) HeadFusion: 360° Head Pose Tracking Combining 3D Morphable Model and 3D Reconstruction. *IEEE transactions on pattern analysis and machine intelligence*, 40(11), 2653-2667.

Yuan, J., Cai, J., Zhang, X., Sun, Q., Sun, F., & Zhu, W. (2019) Fusing Skeleton Recognition With Face-TLD for Human Following of Mobile Service Robots. *IEEE Transactions on Systems, Man, and Cybernetics: Systems.*

Zelinsky, A. (1992). A mobile robot navigation exploration algorithm. *IEEE Transactions of Robotics and Automation*, 8(6), 707-717.

Zanlungo, F., Ikeda, T., & Kanda, T. (2011). Social force model with explicit collision prediction. *EPL (Europhysics Letters)*, 93(6), 68005.

Zhang, D., Xie, Z., Li, P., Yu, J., & Chen, X. (2015, August). Real-time navigation in dynamic human environments using optimal reciprocal collision avoidance. In *2015 IEEE International Conference on Mechatronics and Automation (ICMA)* (pp. 2232-2237). IEEE.

Zhang, K., Zhang, Z., Li, Z., & Qiao, Y. (2016) Joint face detection and alignment using multitask cascaded convolutional networks. *IEEE Signal Processing Letters*, 23(10), 1499-1503.

Zhang, T., Liu, S., Xu, C., Liu, B., & Yang, M. H. (2017) Correlation particle filter for visual tracking. *IEEE Transactions on Image Processing*, 27(6), 2676-2687.

Annexure A

Distance measurement

In Chapter 4, a BSW (Behavioural Stop & Wait machine) was designed. The aim of the BSW machine was to make the robot maintain the same distances that a guide would typically maintain with the visitors. However, the robot forms a reasonably different social class and therefore it is not necessary that the visitors feel comfortable at the same distance with the robot as they would do with the human guide. Therefore, the first aim of the experiments was to find out the distance that the robot should make with the visitors. The problem with the specific work is that the robot is guiding a group and not a single visitor. Therefore, even if a visitor finds some distance comforting with the robot, it is not necessary that the same distance shall be actually possible considering the other members of the group. The second aim of the social experiments is to find out how do the social distances change for small to a fairly large group size. Initially, the face of the human is detected via a face detection algorithm and a rectangular box has been drawn on the face. The height and width of the face is computed by this rectangular box. When the human and robot are very close to each other, then the height and width of the face will be very large. It will be decreasing as the distance increases between the human and robot. When the robot is navigating a group of humans, it is necessary to maintain some specific distance between the human and the robot so that every human can be detected. Here comfort factor is used to find that specific distance. Experiments were carried out to predict the specific distance which will be followed by the robot and human when the robot is navigating all visitors in a group. These experiments have been summarised in Tables A.1 to A.7. A comfort factor of 1 indicates that the visitors were happy with the distance between themselves and the robot, while a comfort factor of 0 means that the visitors believe that the robot is too far off or near. The robot should keep distances within the range of values observed to be comfortable.

Annexure A.1: Distance Measurements Group 2: Computation of width and height of visitors in a group at different distance from the robot

Scenario ID	Image taken by the camera	Number of visitors (actual)	Number of visitors (predicted)	Visitor ID	Weight of face	Height of face	Distance from robot	Comfort factor
1		1	1	1	16	21	2.93	1
2		6	4	1	16	21	2.96	0
				2	14	18	3.39	
				3	17	20	2.95	
				4	18	24	2.70	
3		6	4	1	14	17	3.36	0
				2	17	21	2.95	
				3	14	19	3.96	
				4	21	27	2.70	
4		6	5	1	15	18	3.36	0
				2	18	23	2.95	
				3	17	21	3.18	
				4	20	26	2.65	
				5	29	34	2.10	
5		6	5	1	20	23	2.95	0
				2	20	25	2.65	
				3	16	21	3.18	
				4	20	27	2.63	
				5	22	26	2.34	
6		5	5	1	17	22	2.95	0
				2	16	22	3.20	
				3	13	17	3.97	
				4	16	19	3.58	
				5	19	23	2.95	
7		5	5	1	16	21	2.95	1
				2	16	20	3.05	
				3	15	19	3.36	
				4	15	17	3.45	
				5	18	22	2.98	
8		6	5	1	17	20	3.10	0
				2	16	21	3.18	

				3	16	20	3.36	
				4	15	20	3.28	
				5	18	21	2.91	
9		6	6	1	25	31	2.34	0
				2	22	25	2.45	
				3	16	20	2.58	
				4	16	20	2.71	
				5	14	18	3.36	
				6	16	21	2.89	
10		6	5	1	25	30	2.34	0
				2	17	22	2.81	
				3	17	21	2.96	
				4	16	20	3.36	
				5	19	23	2.61	

Annexure A.2: Distance Measurements Group 3. Computation of width and height of visitors in a group at different distance from the robot

Scenario ID	Image taken by the camera	Number of visitors (actual)	Number of visitors (predicted)	Visitor ID	Weight of face	Height of face	Distance from robot	Comfort factor
1		5	4	1	23	29	2.34	0
				2	19	23	2.95	
				3	18	21	3.18	
				4	29	28	2.1	
2		5	5	1	21	26	2.65	1
				2	16	21	3.36	
				3	18	22	2.95	
				4	17	21	3.18	
				5	26	30	2.34	
3		5	5	1	18	22	3.18	1
				2	24	28	2.33	
				3	19	23	2.81	
				4	20	23	2.65	
				5	26	29	2.09	
4		5	5	1	18	25	3.21	1
				2	39	37	1.37	
				3	16	23	2.95	

				4	19	24	3.18	
				5	24	31	2.34	
5		5	3	1	19	24	2.95	0
				2	39	47	1.37	
				3	25	29	2.35	
6		5	5	1	21	27	2.69	1
				2	39	49	1.37	
				3	19	24	2.95	
				4	16	21	3.17	
				5	22	27	2.34	
7		5	5	1	21	25	2.71	1
				2	37	47	1.36	
				3	20	23	2.94	
				4	17	20	3.18	
				5	21	27	2.63	
8		4	4	1	22	26	2.65	1
				2	18	24	2.95	
				3	16	21	3.18	
				4	22	27	2.65	
9		4	4	1	17	21	3.19	1
				2	18	24	2.93	
				3	18	20	2.98	
				4	21	28	2.65	
10		4	3	1	19	23	2.95	0
				2	16	22	3.20	
				3	21	27	2.66	

Annexure A.3: Distance Measurements Group 4: Computation of width and height of visitors in a group at different distance from the robot

Scenario ID	Image taken by the camera	Number of visitors (actual)	Number of visitors (predicted)	Visitor ID	Weight of face	Height of face	Distance from robot	Comfort factor

1		4	3	1	16	23	2.93	0
				2	16	20	2.97	
				3	23	29	2.68	
2		4	4	1	17	22	2.92	0
				2	14	16	3.89	
				3	17	21	3.01	
				4	21	28	2.66	
3		3	3	1	18	24	2.95	1
				2	17	21	3.18	
				3	21	17	2.55	
4		3	3	1	18	23	2.95	1
				2	17	22	3.18	
				3	21	26	2.65	
5		3	3	1	15	19	3.36	1
				2	15	18	3.36	
				3	19	24	2.65	
6		3	3	1	13	16	3.97	0
				2	13	16	4.00	
				3	17	24	2.65	
7		3	2	1	12	16	3.97	0
				2	22	27	2.60	
8		3	3	1	13	15	3.97	0
				2	14	17	3.58	
				3	23	27	2.65	
9		4	4	1	19	24	2.94	1

				2	15	20	3.36	
				3	23	28	2.34	
				4	25	27	2.65	

Annexure A.4: Distance Measurements Group 5: Computation of width and height of visitors in a group at different distance from the robot

Scenario ID	**Image taken by the camera**	**Number of visitors (actual)**	**Number of visitors (predicted)**	**Visitor ID**	**Weight of face**	**Height of face**	**Distance from robot**	**Comfort factor**
1		4	4	1	18	24	2.95	1
				2	15	19	3.36	
				3	22	27	2.65	
				4	33	39	1.81	
2		4	4	1	19	24	2.95	1
				2	34	41	1.55	
				3	19	19	3.11	
				4	30	38	1.81	
3		4	4	1	45	56	1.41	0
				2	35	42	1.38	
				3	23	31	2.70	
				4	28	34	1.83	
4		5	5	1	47	57	1.41	0
				2	36	41	1.36	
				3	24	30	2.35	
				4	31	39	1.81	
				5	27	33	2.10	
5		4	3	1	19	22	2.98	0
				2	20	26	2.51	
				3	31	38	1.81	
6		4	4	1	20	24	2.98	1
				2	21	25	2.51	
				3	16	20	3.36	
				4	30	37	1.81	
7		4	4	1	18	23	2.78	0
				2	12	16	4.21	
				3	18	22	2.95	
				4	31	38	1.81	
8		4	4	1	18	24	1.78	0
				2	12	15	4.21	

	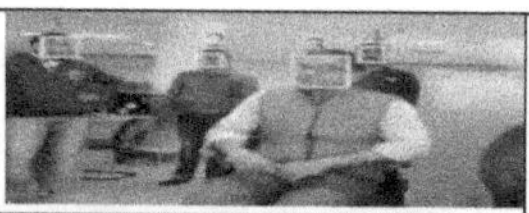			3	30	38	1.81	
				4	14	16	3.36	
9		4	4	1	18	22	1.78	0
				2	12	14	4.21	
				3	14	18	3.21	
				4	13	18	3.36	
10		4	4	1	21	25	2.65	0
				2	12	15	4.21	
				3	22	28	2.50	
				4	15	18	3.36	
11		4	4	1	21	27	2.65	0
				2	16	18	3.18	
				3	22	28	2.50	
				4	14	17	3.36	

Annexure A.5: Distance Measurements Group 6: Computation of width and height of visitors in a group at different distance from the robot

Scenario ID	Image taken by the camera	Number of visitors (actual)	Number of visitors (predicted)	Visitor ID	Weight of face	Height of face	Distance from robot	Comfort factor
1		4	4	1	20	25	2.65	0
				2	12	14	4.10	
				3	23	29	2.60	
				4	15	18	3.36	
2		4	4	1	20	25	2.65	0
				2	11	14	4.22	
				3	21	27	2.53	
				4	15	17	3.40	
3		4	4	1	20	25	2.51	0
				2	15	17	3.36	
				3	20	26	2.53	
				4	13	17	3.42	
4		4	4	1	41	51	1.27	0
				2	28	33	2.10	
				3	45	52	1.16	
				4	67	82	0.92	
5		5	5	1	29	35	1.80	0
				2	43	52	1.12	

				3	40	51	1.31	
				4	38	47	1.37	
				5	66	79	0.92	
6		4	4	1	26	31	2.36	0
				2	36	44	1.55	
				3	67	80	0.87	
				4	29	36	1.79	
7		3	3	1	23	29	2.35	1
				2	20	25	2.65	
				3	20	24	2.95	
8		4	4	1	24	30	2.34	1
				2	22	26	2.74	
				3	22	26	2.65	
				4	18	22	2.95	
9		5	5	1	22	29	2.34	1
				2	23	26	2.65	
				3	20	24	2.60	
				4	18	21	2.95	
				5	16	20	3.18	
10		6	6	1	19	24	2.95	1
				2	16	21	3.25	
				3	21	26	2.59	
				4	18	21	3.17	
				5	25	30	2.31	
				6	22	27	2.66	

Annexure A.6: Distance Measurements Group 7: Computation of width and height of visitors in a group at different distance from the robot

Scenario ID	Image taken by the camera	Number of visitors (actual)	Number of visitors (predicted)	Visitor ID	Weight of face	Height of face	Distance from robot	Comfort factor
1		5	5	1	24	29	2.42	1
				2	20	26	2.75	
				3	20	24	2.47	
				4	18	23	2.82	

				5	21	25	2.49	
2		5	5	1	25	30	2.42	1
				2	20	26	2.75	
				3	37	44	1.40	
				4	16	21	2.82	
				5	22	27	2.49	
3		5	4	1	15	19	3.66	0
				2	16	20	3.36	
				3	14	19	3.89	
				4	22	26	2.75	
4		6	5	1	22	27	2.72	0
				2	15	19	3.66	
				3	19	26	2.89	
				4	15	20	3.36	
				5	17	23	2.75	
5		6	4	1	14	19	3.89	0
				2	14	22	2.75	
				3	16	19	3.36	
				4	20	27	2.68	
6		5	3	1	14	18	3.89	0
				2	13	17	3.96	
				3	14	18	3.36	
7		4	2	1	20	27	2.67	1

				2	19	24	2.75	
8		4	3	1	21	26	2.67	0
				2	27	35	1.67	
				3	17	22	3.11	
9		3	3	1	17	23	3.15	1
				2	16	21	3.15	
				3	17	21	3.15	
10		3	3	1	14	19	3.81	0
				2	22	27	2.14	
				3	14	18	3.81	
11		3	3	1	29	33	2.14	1
				2	28	32	2.14	
				3	23	33	2.14	

Annexure A.7: Distance Measurements Group 8: Computation of visitors in a group at different distance from the robot

Scenario ID	Image taken by the camera	Number of visitors (actual)	Number of visitors (predicted)	Visitor ID	Weight of face	Height of face	Distance from robot	Comfort factor
1		2	2	1	14	19	3.81	0
				2	27	31	2.14	

2		2	2	1	14	19	3.81	0
				2	14	18	3.81	
3		2	2	1	25	31	2.42	1
				2	24	30	2.42	
4		1	1	1	24	31	2.42	1
5		4	3	1	25	30	2.57	0
				2	15	21	3.54	
				3	21	26	2.59	
6		4	3	1	22	26	2.57	0
				2	29	37	1.78	
				3	21	25	2.59	

7		3	2	1	19	24	2.45	1
				2	21	25	2.59	
8		2	2	1	19	24	2.44	1
				2	19	23	2.88	
9		2	2	1	18	23	2.88	1
				2	19	23	2.88	
10		1	1	1	16	19	3.05	1

11		1	1	1	18	22	2.88	1

Chapter 6

Conclusion and Future work

6.1 Conclusions

In general, humans display several behaviours in a group when they are following a leader robot such as swapping their position, overtaking each other, do not follow the leader robot, or overtake the leader robot. Due to these behaviours tracking the humans continuously from a mobile robot base is a challenging problem. The problem is amplified because of a highly limited field of view of the onboard vision system. Since this thesis is based on social navigation and tracking therefore the robot must be capable of handling these behaviours. Moreover, the robot navigation for a group of visitors in indoor scenario is a related challenge, where the robot should adapt as per the different behaviours of the humans. The application includes robotic guides in museums, shopping centres and other places. The robot during navigation in such places will also witness other people who are not a part of the group and thus act as an obstacle. Hence motion planning algorithm is very important for the robot movement. The social robot should use the human conventions and social etiquettes for the motion planning to avoid the obstacles and achieve the goal.

In this thesis, a particle filter-based algorithm was presented to track multiple humans in a group with multi-behaviours and human following the robot behaviour in Chapters 3 and 4. A social robot tour guide application with Behavioural Stop & Wait Machine and Behavioural Finite State Social Machine was also presented in Chapter 5. A social robot motion planning with social cues inspired from the Indian pedestrian traffic rule was presented in Chapter 6. The thesis develops the algorithms for a social robot used for the navigation of the humans and the humans have to follow the robot for their navigation. In a group, different behaviours of the human were accounted for and understood by the robot to perform successful navigation and tracking. Apart from this when robot has to take a sharp turn at a corner, due to lack of space the humans cannot place themselves behind the robot. There is a loss of visibility of the humans. The different modules accounted for this uncertainty and neither made the robot slow down nor deleted the tracks of the tracking algorithm. Robot motion planning was undertaken using a human convention to achieve the goal and avoiding the obstacle. The robot typically preferred to avoid the oncoming people from the left, while a right-side avoidance could still be chosen if the person to avoid was sufficiently in the opposite side. A large part of the research borrows ideas from general human behaviour to carry out motion planning with navigation in a human-like behaviour; and social tracking based on the navigation strategy of the people in a group behind the robot. The research focuses on the case when the robot takes a movement at a corner where space is not available behind the robot for the humans to follow. The Behavioural Finite State Social Machine was used for the navigation and Social Particle Filter model was used for the tracking.

Social 3D tracking was performed for multi-human tracking in a group while the humans perform different social behaviours such as swapping, overtaking, getting missed, etc. 'Human Following Robot' behaviour was developed for a single person being tracked by the robot, while the person follows the robot. This was extended to a multi-behaviour social particle filter which handled all these behaviours of humans in a group. The algorithm used a social potential field to predict the motion of the humans in a group as a motion model. This model was capable of executing 3D tracking with a low-cost budget camera and also produced less errors in tracking as compared to the other methods. The proposed model gave a very small tracking error of 425.26 mm as compared to the other approaches (Random Particle Filter, Tracking in Image, LSTM with re-initialization, and LSTM without re-initialization). The position is needed to enable a socially pleasing human robot interaction. The error is small enough for the robot to approximately look and interact with the visitor. The proposed model was also able to handle different erratic behaviours. Testing was carried out by using 4 occurrences of overtaking, 9 occurrences of swapping positions, 5 occurrences of visitors following another human, 9 occurrences of visitors leaving the group and later re-joining the group. The proposed model in nearly all cases correctly detected a person leaving the group and the track of the person was deleted. In a few cases there were some false positives because the humans cannot place themselves behind the robot due to a limited space and a limited field of view of the camera. A visitor has no feedback about the visibility in the camera and therefore the person cannot adjust the position to cooperative.

Similarly, the proposed approach correctly and promptly re-initialized the track once the person re-joined the group.

The robot was deployed as a guide to take visitors on a tour using a Behavioural Stop & Wait (BSW) machine that made the robot pause the tour if the visitors are not detected by the robot. When experiments were performed it was observed that the robot unnecessary stopped due to lightning condition change and detection errors. The robot produced jerks on stopping. A total of 6 false positives were observed, where the robot stopped even though the visitors were present. The approach was extended by using a Behaviour Finite State Social Machine (BFSSM). Here the artificial potential field was utilised for motion planning. A speed controller module was responsible for handling the velocity of the robot and also gave an acknowledgement to the missing visitors expecting them to re-join the group. The robot slowed down on not being able to detect the visitors as a soft signal to the visitors to catch up. Further, this machine also anticipated a lack of visibility at the corner and therefore false positives were rapidly decreased. The robot was also skilful at detecting corner points during the whole journey time. Using the BSW if the robot turns at a corner, its rear camera may be facing a wall and the visitors will have to make themselves visible that may not be possible, making the robot stuck. In that case the robot cannot complete the tour. In the tested scenarios, the visitors faced this situation, and the robot took an additional time of 21-52 seconds in different scenarios to detect the visitors. The robot saved this time with the BFSSM that reduced the average journey time by 27.90% as compared to BSW. BFSSM generated zero false positives and therefore the journey ends in a small time of 131.17 seconds.

The robot demonstrated intelligent behaviour inspired from the pedestrian and the Indian traffic rules for successfully achieving its goal. Social cues were used for motion planning of the robot which were injected into the artificial potential field. The resultant algorithm was named as the social potential filed with social cues (SPF-SC). The same social cues were also injected into a geometric planner, named as Follow the Gap with Social Cues (FTG-SC). A fusion of the potential field and the geometric planner was made. The hybrid approach improved the average clearance by 28.87% in overall scenarios that increased the safety of the robot. The proposed hybrid approach maintained the largest average clearance of 0.67 meters, followed by 0.58 meters maintained by SPF-SC as the second-best algorithm. Apart from that, two more approaches were compared with the proposed method. The artificial potential field (APF) gave the third highest clearance of 0.52 meters, while FTG-SC algorithm gave the fourth highest average clearance of

0.49 meters. The average penalty for not maintaining the required social distance by the proposed approach was nearly 0 in all scenarios, significantly lower than the other algorithms. A small penalty still exists because in some cases the robot has to pass through narrow corridors amidst obstacles. The hybrid approach obtained a very less penalty of 3.01, followed by the second smallest penalty of 20.32 obtained by SPF-SC. The two other approaches FTG-SC and APF have a larger penalty of 23.82 and 52.69 respectively. The hybrid approach reduced the average time by 14.29% to achieve the goal as compared to the baseline approaches. The hybrid approach took an average time duration of 41.09 seconds while SPF-SC took 41.3 seconds. The other approaches FTG-SC and APF took more than 10 additional seconds with a timing of 51.3 seconds and 52.99 seconds respectively as compared to proposed hybrid method. Although, this is a very less time difference, but it is obtained in a very small environment for a single goal. If the experiment is performed in a large workspace with multiple goals, this time difference will increase rapidly demonstrating the utility of the hybrid approach.

6.2 Scope for future work

One of the major limitations currently is in benchmarking tracking. Currently, while the experiments were carried out under realistic settings with the complete robotic setup including humans, the benchmarking was carried out using simulators to get the ground truth. Humans may not have socially simulated the robots. The experiments need to be done using external trackers to prepare the ground truth. The major application was enabling the robot to take an informed decision based on the tracked positions of the people. Using the tracker uncertainty as an input in the planning algorithm of the robot needs to be explored.

In the future the proposed approach will also be modified to select the largest clearance as an enhancement of the approach. This can be carried out both at the global and the local level. So, even if multiple paths exit, then the robot will choose the path which has a large space or gap between the obstacles or boundaries of the workspace. In this scheme the possibility of the robot getting close to the obstacle or workspace boundary will be resolved enabling people to easily come within the field of view of the camera for a better social experience.

This thesis considered a few challenging scenarios for testing. However, in real-life many things can happen such as if people go into a shopping centre, the scenario and the people's behaviour will be very different, while on an academic campus or pilgrimage this behaviour and situation will be different. In the current scenario, the experiments were performed in the Centre of Intelligent Robotics at IIIT Allahabad (Prayagraj). In the future, other scenarios will be explored such as warehouses, academic campus, shopping centres, etc. to examine and evaluate the different constraints and its requirements related to the proposed model.

The thesis presented numerous social behaviours in the navigation and tracking of the robot. A typical social robot also interacts with humans, which was not the current scope of the work. While navigating, asking for space from the human and giving the human space on demand are social etiquettes of navigation which were not explored in the current work. The extension of the work for different kinds of robots with different capabilities and for different applications needs to be done. Possibilities for adding an active human-robot-interaction during navigation also needs to be studied. Furthermore, the behaviour of the robot needs to be enhanced by studying more social behaviours between tour guides and people; and transferring the same into the robots. The approach needs to be extended to large tours in a multi-robotic setting. Incorporating the desire, mood and preference by looking at the observed behaviour of the human also needs to be looked in the future. A challenge is that tracking from a mobile base need to be very accurate as tracking errors can misclassify the preferences, mood or speed parameters.

The thesis identified some social behaviour and displayed the performance of the different algorithms reproducing those behaviours. However, the identified behaviour may not be exhaustive and may further vary with different kinds of people. The task of learning and classification of human behaviour and the variance of behaviour in different contexts, scenarios and with different people needs to be studied in the future. Further, the assessment of such trajectories to gather examples of social etiquette and to thereafter inject the same in the motion behaviour of the robot is the open topic of research for the future.

In this thesis, humans are detected by their face but in the current situation their faces are covered with a mask due to the COVID-19 pandemic. Colour or dress-based encoding for the human detection and tracking may be tried in the future. Currently the visit places are well localised, and the robot uses only the coordinate system for navigation. Guides can often reorder places or stand at different places depending upon the crowd which is a capability that may be added in the future. In the future, the robot may be asked to identify the visiting place using deep learning and scene recognition. The same may also be used for an active localisation in crowded scenarios.

In the thesis, people are observed and tracked by their faces. The tracker assumes attraction and repulsion between the people, which at equilibrium will lead to patterns formed by the robot and the group of people such that the people are relatively comfortable. The pattern formed is governed by different coefficients of the social force model, that are currently set by intuition and a knowledge of the equilibrium distance measured by some social experiments. These coefficients and hence the pattern formed depends upon the preferences of the people in the group, culture, and other factors. The tracked position can be used to determine how the people are moving and what pattern they are making. In the future, people can be observed for some time and a new algorithm can be used that will be able to learn the different types of patterns that the people form between themselves. The algorithm will be able to identify which patterns are good for better motion prediction and resultantly, a self-adaptive motion model will be proposed as a future work that will use an unequivocal parameter setting based on the observed pattern formation of the people. In this way the robot may act as an expert guide that adapts its behaviour based on the perceived behaviour of the other people.

www.ingramcontent.com/pod-product-compliance
Lightning Source LLC
LaVergne TN
LVHW010716061126
851857LV00014B/1222
* 9 7 9 8 2 5 6 2 5 4 4 9 0 *